The AJM Guide to
Lost-Wax Casting

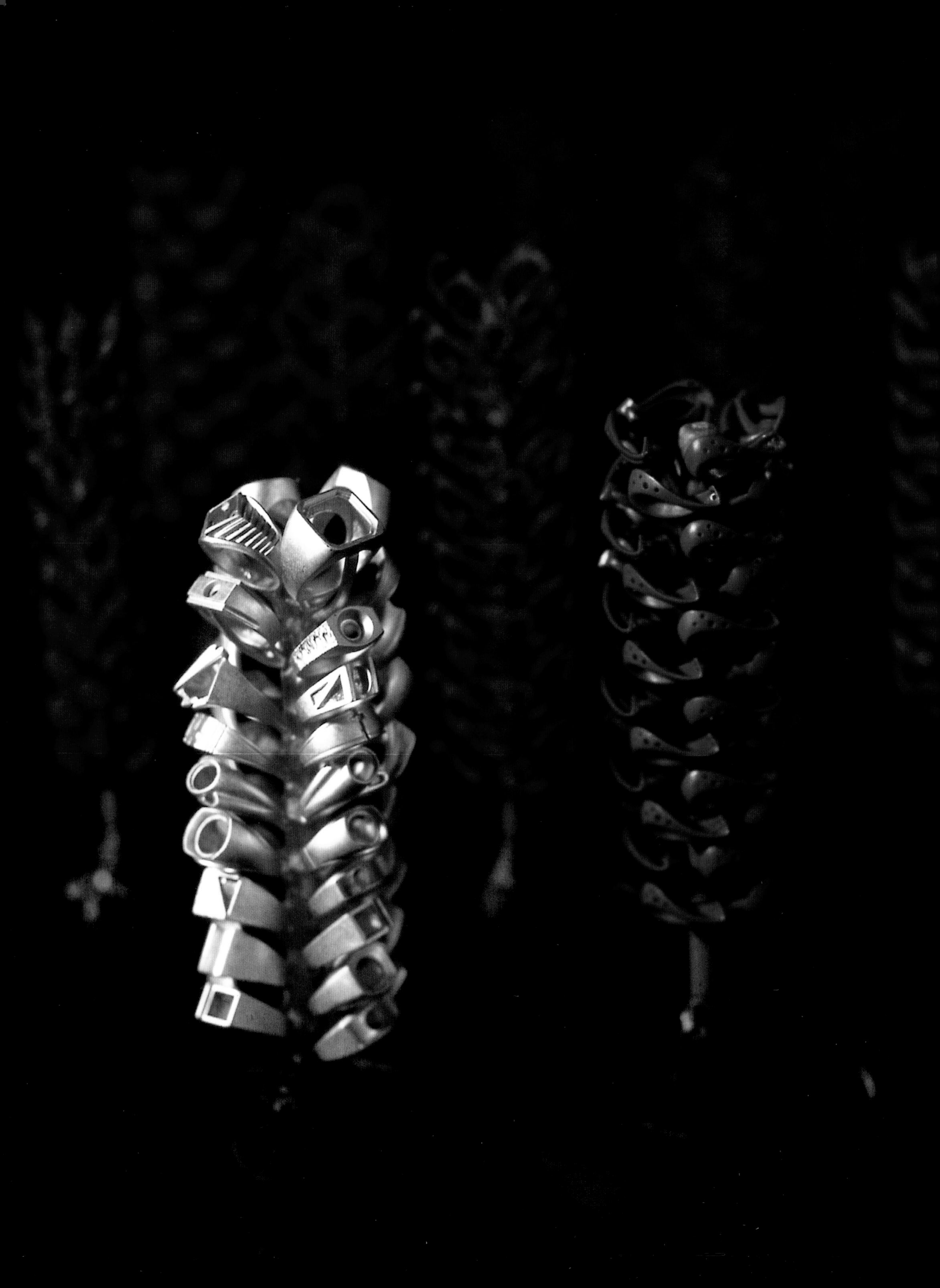

# The *AJM* Guide to Lost-Wax Casting

## Techniques and Tips from Industry Experts

MJSA/AJM Press
A Division of Manufacturing Jewelers & Suppliers of America Inc.

ISBN 0-9713495-2-5
Library of Congress Control Number 2003110728



**Photo Credits**

**Daniel Ballard**, Precious Metals West/Fine Gold, Los Angeles: Page 60.
**Donna Ballard**, West Toluca Lake, California: Page 62.
**Gary Dawson**, Goldworks, Eugene, Oregon: Cover and pages 5, 101, photo (from top) 4; pages 58, 63, 65, 98, 99.
**Valerio Faccenda**: Page 84. Used by permission of Christopher J. Cart.
**Greg Gilman**, Stuller Inc., Lafayette, Louisiana: Cover and pages 5, 101, photos (from top) 1, 2, 3, 5, 6; pages 2, 6, 8, 12, 14, 15, 16, 24, 28, 30, 32, 38, 40, 41, 42, 43, 44, 50, 51, 52, 66, 72, 78, 96.
**Jurgen J. Maerz**, Platinum Guild International USA, Newport Beach, California: Page 89.
**Hubert Schuster**, Jewelry Technology Institute, Vicenza, Italy: Page 92. Used by permission of Eddie Bell.
**John Sciarrino**: Pages 18, 20, 21. Used by permission of Chuck Hunner.
**J. Tyler Teague**, JETT Research, Albuquerque, New Mexico: Pages 34, 35.
**Weston Beamor**, Birmingham, United Kingdom: Pages 80, 81, 82. Used by permission of Christopher J. Cart.

The majority of articles in this book originally appeared, in modified form, in *AJM* Magazine. *AJM* Magazine is published monthly by Manufacturing Jewelers and Suppliers of America Inc. Subscriptions: U.S.: $42USD one year, $75USD two years; Canada and Central America: $54USD one year, $96USD two years; all other countries: $88USD surface, $141USD airmail.

For information about subscribing to *AJM*, call 1-888-438-6572 (in U.S. and Canada only) or 1-401-274-3840; fax 1-401-274-0265; e-mail *ajm@ajm-magazine.com*; Web site *www.ajm-magazine.com*.



Book design by Jesse Snyder, Skies America, Beaverton, Oregon
Printed in Hong Kong

*The* AJM *Guide to Lost-Wax Casting* is published by MJSA/AJM Press, 45 Royal Little Drive, Providence, RI 02904; 1-800-444-6572 (U.S. only) or 1-401-274-3840; fax 1-401-274-0265; e-mail: *mjsa@mjsainc.com*; Web site: *www.mjsainc.com*.

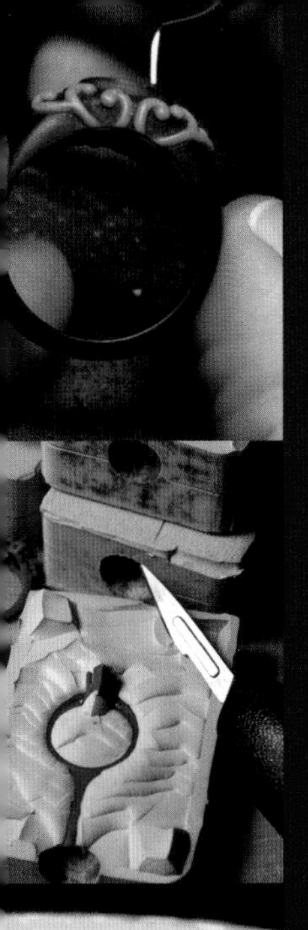

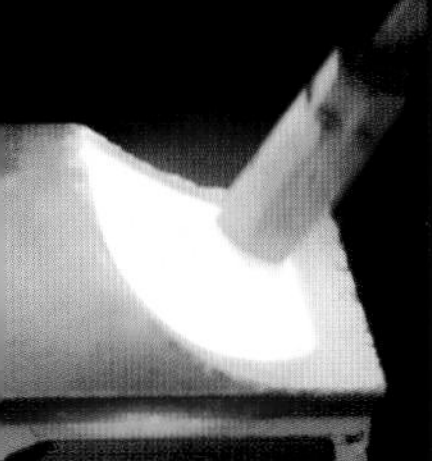

# Table of Contents



## Casting Resources

# Introduction

Ask any caster to name the one crucial ingredient for success in lost-wax casting, and more times than not you'll get in response a little chuckle (if not a full-fledged belly laugh). That's because, in casting, the final product is only as great as the sum of its parts.

A smooth, crisp, porosity-free casting results not from the caster doing any one thing right, but from careful attention to a plethora of details. Does the master model have a suitable surface finish? Has the mold been adequately packed? Does the sprue network ensure sufficient metal flow? Does the investment have the necessary consistency, and has an appropriate time-temperature sequence been selected for the burnout stage? Is the metal being cast at the correct temperature? Has the best method been chosen for breaking out the castings? These are just some of the questions that casters must answer; even slight deviations from the proper procedures can, over the course of the process, add up to disaster.

During its existence, *AJM* Magazine has published numerous articles that attempt to answer these questions. As the flagship publication of Manufacturing Jewelers and Suppliers of America, *AJM* shares the ultimate goal of its parent association: to advance the jewelry industry. And to advance the practice of lost-wax casting, the magazine has offered in-depth technical and business information supplied by the industry's leading experts. It has addressed every area of the process—from developing the master model to breaking out the castings—all in the hope of making casters' lives a little easier, and ensuring consistent quality in the final product.

To further that aim, we decided to take our efforts one step further: What if casters, rather than relying on voluminous back issues, could turn to one publication that presented this information in a concise, easy-to-use format?

The result is the book you now hold in your hands: *The* AJM *Guide to Lost-Wax Casting.* It features some of the most notable articles on lost-wax casting that *AJM* has published over the past several years (as well as a few new pieces); all have been arranged so that the reader can go through the entire casting process, from start to finish.

Throughout these pages, you'll find an abundance of best practices, handy tips, and troubleshooting techniques, as well as insights into a few of the scientific principles behind the processes. You'll read about some of the latest innovations, and about the pioneers who are continuing to further the art of casting. And you'll find a variety of resources to ensure you have access to the needed tools, materials, and services.

As in casting itself, this book is the sum of many parts—in this case, the efforts of many contributors. We would like to thank them all for graciously contributing their time and expertise, and for sharing their knowledge with the rest of the industry. By doing so, they ensure that the art of lost-wax casting not only endures, but advances—for in the end, it's knowledge itself that is the crucial ingredient for true casting success.

The *AJM* Editors
August 2003

# A Model Approach

## PLANNING AND CREATING THE MASTER

BY GREGG TODD

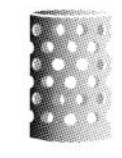

A question I often hear when making new master models is: Where do you get your ideas? I typically answer that I get them from all around, from nature or the like, and the inquirer is usually placated. But the one question I have never been asked—and the one that would generate the most panic—is: How close is the master to the design you had in mind?

Making a master model look exactly like the intended design is one of the most difficult challenges faced by the model maker. For one thing, the master has to be made larger than the intended finished piece to accommodate shrinkage. In addition, things look quite a bit different on paper than they do in metal: I have seen delicate designs on paper become nearly indistinct surface markings on a casting when accurately reproduced.

Then there's the challenge of figuring out what the designer wants—which sometimes requires considerable interpretation. I have seen instructions that say, "I want the piece to look exactly like this, or how you think it will look best." Talk about confusing.

Model making can be viewed as an exercise in intelligent compromise. The model maker must find an acceptable middle ground between what the designer wants and what can be done within the limits of casting. And the best results can be achieved when all parties involved understand those limitations.

## Planning for Shrinkage

The first factor to consider when creating a master model is that all reproduction methods result in shrinkage. The degree to which shrinkage will affect the final product depends on the piece. For example, if the design is a freeform mounting without seats for stones, shrinkage may not be an issue. On the other hand, if there are extremely tight tolerances for set stones, shrinkage becomes very important.

Most models are made in one of three ways: direct fabrication, casting, or a blend of both. No matter what method is used, there will be some shrinkage from the master model to the finished piece. The amount will depend on a variety of factors, but it usually ranges between 6 and 12 percent.

As a general rule, there is less shrinkage to account for when the model is fabricated than when it is cast. This is true because there is an additional step where shrinkage can occur in casting: the wax model. Once the wax master has been cast into a metal master, the amount of shrinkage from that point on will be the same as with a fabricated master. This additional shrinkage must be considered when carving an original wax, as well as when the master is being made through a blend of cast parts and fabricated parts.

Although the rate of shrinkage will be the same throughout the piece, the appearance can vary. Let's say, for example, the piece shrinks 8 percent. If one part is 10 mm and another is 2 mm, it will appear that the 10 mm part had considerable shrinkage, while the 2 mm part had little to none. In mathematical terms, the 10 mm part shrank a noticeable 0.8 mm, while the 2 mm part shrank a nearly invisible 0.016 mm.

There is no single set value for shrinkage because the exact amount of shrinkage depends on your specific processes: whether a metal mold or a rubber mold is used, the type of wax used for producing the pattern, the metal used in the casting process, the temperatures of the mold and the metal, and the method of casting. You will have to determine the most accurate value for your work through trial and error. One way to do this quickly and accurately is to cast some wax wire of a known size. Use the same type of wax used to make the pattern. Measure the diameter and the length of the wire, and record the size before and after casting. Next, calculate the area before and after, using this simple formula: ($\pi r^2$ x length). You can then calculate the percentage of shrinkage with this formula: ([before area - after area] ÷ before area).

You will have to repeat these steps to account for your entire shrinkage factor from master model to finished casting. Minor variations in processes from one piece to another will yield slightly different results, but you should be in the same ballpark.

## Preparing Stone Seats

Model makers are frequently asked to build a model to accept a specific stone size, in reference either to weight or to dimension. In these cases, shrinkage can cause serious problems, so it is often advisable to allow for slightly more shrinkage than anticipated.

When the stones are relatively large (3 mm and up), it is particularly important to allow for more shrinkage. For example, when preparing a master model for a 3 mm stone setting, it's a good idea to add 0.24 mm (which is 8 percent of 3 mm) for a total size of 3.24 mm. If the total shrinkage turned out to be 12 percent, for example, you would need the setting to measure 3.36 mm, a difference of just 0.12 mm. The setting process will probably accommodate this small difference.

Other factors should also be considered. First, unless calibrated stones are used every time, the intended stone size may not be exact. A 3 mm seat may be expected to hold stones ranging from 2.8 mm to 3.2 mm, if the stones are not adjacent to one another. For stones that are adjacent to one another, the range may narrow slightly to between 2.8 mm and 3.1 mm. If the seat is, in this case, made exactly to size, it would be unable to accommodate any stone larger than 3 mm without overlapping.

I generally allow for more shrinkage when designing seats for adjacent stones, usually 12 to 15 percent. If the requested stone size is 3 mm, this shrinkage calculation will result in seats of 3.36 mm (allowing for 12 percent shrinkage) to 3.45 mm (allowing for 15 percent shrinkage). If the actual shrinkage proves to be 8 percent, the seats end up between 3.1 mm and 3.17 mm, which again can be included in the stone setting process.

When I am confident that I know what my actual shrinkage rate will be, I can add that amount to the maximum seat size. In the previous example of the adjacent stones, the largest finished seat size desirable would measure 3.1 mm. If my shrinkage factor were 8 percent, the seat in the master model would have to be 3.35 mm; this would allow the cast mounting to accept the desired range of stone sizes.

Occasionally, a master model is made to such precise standards that unless the stones are carefully calibrated, they simply will not fit. A far better approach is to leave a little extra room for the stones, allowing for normal variations in size.

## Finishing the Master

I once believed that the surface of the model did not have to be any finer than the investment, because the investment controlled the surface quality of the casting. Then I was re-educated. The model isn't used to make the casting: The model is used to make a cavity in an injection mold to produce a pattern. As a result, the surface quality of the model has a direct impact on the quality of the pattern. And since a casting cannot be any better than the pattern—not to mention that rough surfaces on a pattern can contribute to turbulence during casting—surface finish is an important issue in the master model.

Waxes should be cleaned and shined with solvent, and metal should be brightly polished in all areas: poor finish increases the surface resistance between the mold and injection wax. The greater the resistance, the more likely a distorted or broken pattern will result.

In addition, if it is difficult to remove patterns from a mold, the mold will be worked harder, shortening its lifetime. Fine parts of the mold are likely to tear from repeated stretching, as well as suffer a loss of memory that results in alignment problems.

It is sometimes thought that if the surface of the finished casting is intended to have a matte finish, the model doesn't need to be polished. This argument ignores the above points. In addition, because surface finishes are one of the last things applied to the casting during the finishing process, they do not have to be present when the pattern is made. Even if you intend to have a matte finish, you should brightly polish the master.

## Factors Often Forgotten

Other factors to consider when planning master models are the effects of the finishing process on the models, the right time to add details to them, and the type of metal that the model maker should use when building it.

While shrinkage is the major concern in figuring out the correct size for the master model, the model maker must remember that polishing also reduces the outside dimensions. Unlike shrinkage, this loss is seldom uniform. There is more reduction on the extremities, since these are in greater contact with polishing wheels. Interior corners also have higher rates of metal removal, due to the focusing effect of brushes and buffs. When possible, these areas should be made slightly thicker.

Many model makers also fail to fully appreciate the amount of detail that can be achieved in the casting process. Most model makers feel that the best detail can be produced in metal. However, while wax does not offer the same resistance as metal, it accepts details quicker, and errors are easily corrected.

The quality of detail will vary with the skill level of the wax carver; but whatever your skill level may be, it can be developed through practice. A skilled wax carver can complete 90 to 95 percent of the detail in wax and the remaining 5 to 10 percent in metal. This allows faster completion of the master.

Finally, the model maker must consider the metal to be used in creating the master model. The most popular metal for models is sterling silver. It provides an excellent compromise between ductility, durability, and malleability. It also cuts easily, can be soldered, and maintains detail well.

But in some instances, such as when a very large or very delicate piece is created, silver may be too malleable to hold its shape. On these occasions, other metals can be used instead of sterling. For example, fine pierced filigree detail, which can be deformed easily in sterling, becomes much more rigid when made in white gold. (White gold is less malleable and the added alloys are less expensive than yellow gold, making it a good choice when a gold master is required.) The use of gold, rather than silver, will increase the cost of the master, but compared to the cost of repairing it over its useful life, the added expense of the white gold can be negligible.

You can also rhodium plate the silver model, after first plating it with nickel to prevent contamination of the rhodium. The nickel has the added benefit of leveling the surface, and also provides a tough skin over the model that can reduce damage when mold cutting.

Once you've ironed out all of these details, the final point to ensure is that your master model is flexible. It should look like the intended design, while at the same time allowing for the inevitable variations encountered during the production process. So the best answer to the question, How close is this model to the design you had in mind? is probably, close enough.

# Wax Works

## ENSURING A MODEL MEASURES UP

BY GREGG TODD

Of all the details that must be addressed in lost-wax casting—water-powder ratios, burnout temperature ramps, sprue placements, etc., etc.—perhaps the most important is creating the master model. As the first step in the process, it is the root of all success—and all failure. A pattern with overly thin walls, a poorly transcribed design, or badly detailed reliefs will ultimately lead to a poor casting, regardless of how well performed the rest of the operation may be.

Although there are many techniques for developing wax models in preparation for casting, one of the most common is wax carving. With this in mind, I would like to offer a few tips to the intrepid few (well, multitudes) who often find themselves with the challenge of translating an imaginative design into a 3-D wax model.

## Truing the Wax

One often overlooked step in preparing a wax for carving is truing. This step involves the establishment of right-angle intersections at the critical junctions where different surfaces meet, so that all measurements can be referenced. This is very important, since the surfaces of the wax tube can vary: I've seen ring tubes with the center hole out-of-round and misaligned from one end of the tube to the other. I've also seen outer surfaces that have profound sinks and other defects.

The simple slide gauge is one of the best tools available for checking the right angles; that, a file, and a pair of dividers are the only tools required for the process. Place the flat edge of the gauge against the referenced surface, then slide it against the surface to be aligned.

Work your way around the surface, taking care to file the high points until a single flat surface is achieved. (Depending on the severity of the surface condition, you may need to start with a coarse file and transition to a finer cut.) After the surface has been established, the dividers can be used to mark a surface parallel to the first.

Once the wax has been prepared, it's then ready for carving—and, hopefully, the following techniques and tips will make that job a little bit easier.

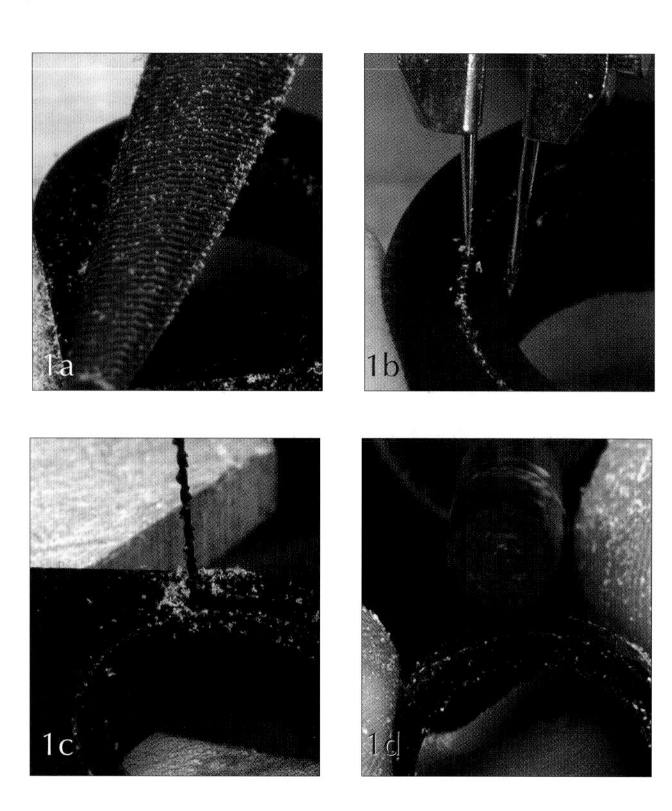


## Shaping the Ring Blank

The creation of a ring always begins with the preparation of the ring blank. Cut a slab from a wax tube that has been trued and is slightly wider than the ring will be. (This assumes, of course, that you are working from a scaled drawing, which is most advisable.) The blank will now be at its maximum dimensions—which means that it will also be at its maximum strength.

Begin sizing the blank to the desired ring size. Slide the wax blank on a ring tube sizing mandrel, then gently turn it against the blade to begin shaving the interior surface of the ring blank.

Since the mandrel is tapered, you should repeat this process from both sides of the blank. This will leave the center slightly thicker than the outer edges, but the wax at the center will be removed later.

Once the wax blank has been sized, file and smooth each side to remove the rough surface produced from sawing the blank to width (Figure 1a).

Using a pair of dividers set to the desired thickness of the ring, scribe a line on both sides of the wax blank, just above the inside diameter (Figure 1b). These will become the layout lines for roughing the band to proper thickness.

Although the blank could be filed to shape, this can be time consuming. Instead, try using a saw to initially reduce the overall dimensions of the blank (Figure 1c). Spiral saw blades are the best choice for quick cutting, but make sure to use slow, deliberate strokes. Rapid sawing often results in the wax melting and rejoining behind the saw blade, so that the parts do not separate.

Begin filing the wax blank to the desired thickness as indicated by the layout lines scribed earlier. A double-ended wax file or rotary file (Figure 1d) are suitable choices for performing this step. Once the thickness has been reduced to the desired size, smooth the outer surface either by filing or by scraping.

Now it's time to remove that extra thickness produced during

the sizing step. Hold the ring tube sizing mandrel so it touches only the center surface of the ring blank (Figure 1e) and carefully shave off the high point running around the center of the band. This process is easier to accomplish if you do not slide the ring up the mandrel to the desired size: Remember, the ring has already been sized, and now you just want to make the interior of the band flat.

## Transferring the Design to the Wax Blank

The technique that I am about to describe falls within that realm of closely guarded secrets that few are allowed to possess. Begin by making three copies of the scaled drawing: one to work with, one to use as a reference, and one just because jewelers are human, and it's always good to have a spare.

Take one of the copies and cut out the design, leaving a little extra around the edges. Lay the copy face down on a soft surface, such as a few napkins. With a wax pen, melt and coat the back of the drawing with sprue wax (Figure 2a). Sprue wax is soft and generally has a lower melting point than most other types of wax, which is critical to the next step.

Hold the wax-coated surface of the drawing tightly against the ring blank and, using the wax pen, iron the drawing to the surface of the ring blank (Figure 2b). As the wax on the paper melts and bonds to the surface of the pattern, it becomes visibly darker. Use this as an indicator to ensure the entire design is bonded to the ring blank.

Transferring the design to the ring blank is now just a matter of tracing the design with a scalpel (Figure 2c). You should always use a new blade for this step; dull blades often tear and shred the paper, and they fail to produce a clear line in the ring blank.

Occasionally a design will pierce through the band. Leaving the paper pattern in place (Figure 2d) while cutting out these areas can be helpful for two reasons: It is much easier to see the layout and stay within the lines, and it helps to prevent chipping along the top edge of the pattern.

When all of the piercing is complete, remove the paper pattern and shape the sides of the ring to the desired contour using files, burrs, scrapers, and gravers. Continue to smooth and refine the surface until it matches the design. Polish the wax model, adding any fine detail that might have been polished out. (For the pictured design, which called for small half-round depressions near the points of the triangles formed between the pierced openings, I used a small ball burr turned between my fingertips.)

Felt points, cotton swabs, and nylon-stocking material work well for polishing. Both the points and the swabs can be used with a wax solvent to smooth and level the surface. Be sure to clean the model after using solvents: They can continue to react with the surface after polishing, as well as cause problems during investing.

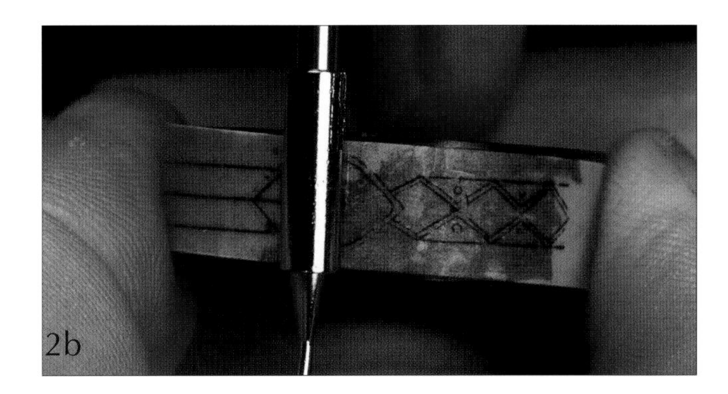
2b

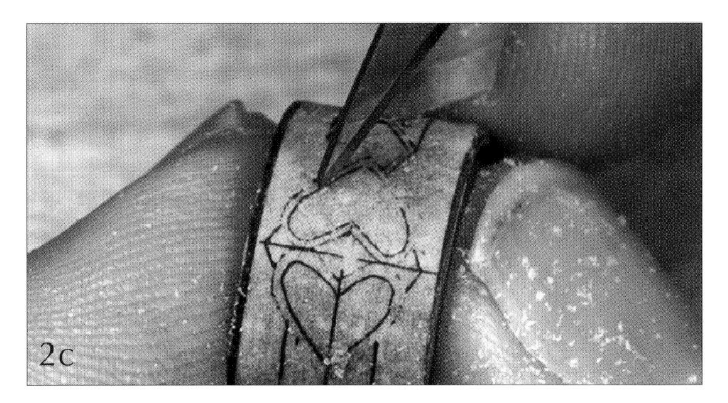
2c

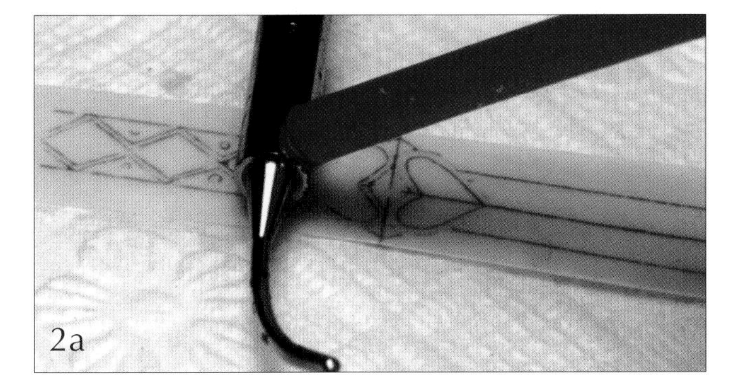
2a

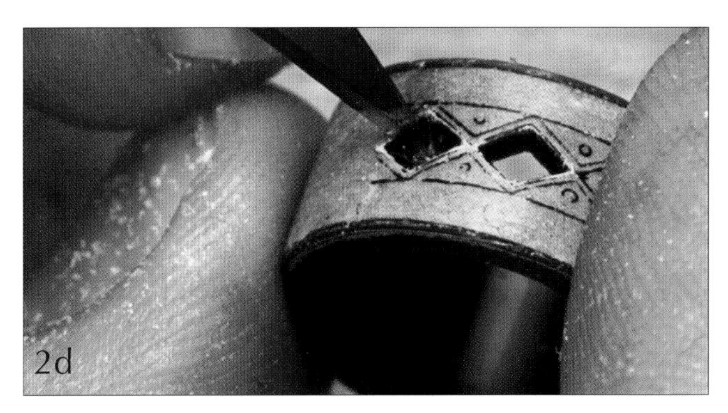
2d

## Creating a Relief Pattern

Carving jewelry with high- and low-level relief presents additional challenges to the wax carver. I prefer to create such a piece by dividing it into two sections: the carved relief pattern and a back plate on which the relief is set. This method not only minimizes shrinkage, but also enables jewelers to cast each piece in a different colored metal.

Begin by preparing a thin plate for the relief from a block of carving wax. Although sheet wax can also be used, I prefer carving wax because of its hardness and resistance to deformation.

After sawing and filing the wax to produce the plate, use the edge of a flat graver (Figure 3a) to remove the file marks and smooth the surface. Once that is done, the pattern can be prepared and transferred in the same manner as described on page 15. Next, with a 4/0 saw blade, begin sawing the outer contour (Figure 3b). The edges of the pattern often begin lifting during the sawing process, so use your wax pen to iron them back down (Figure 3c). Once the sawing is completed, use a scalpel to mark the layout lines in the wax sheet. When the paper pattern is removed, the layout lines should be visible on the surface of the piece. Use gravers and scrapers to first define the layout lines, then to contour the surfaces. The back plate is prepared in the same manner as the original sheet.

3a

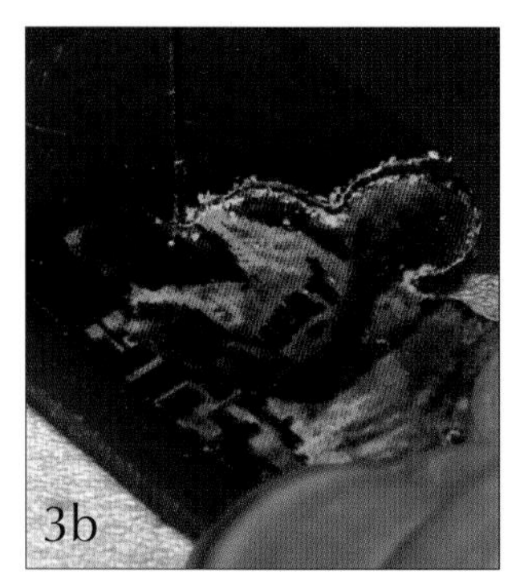
3b

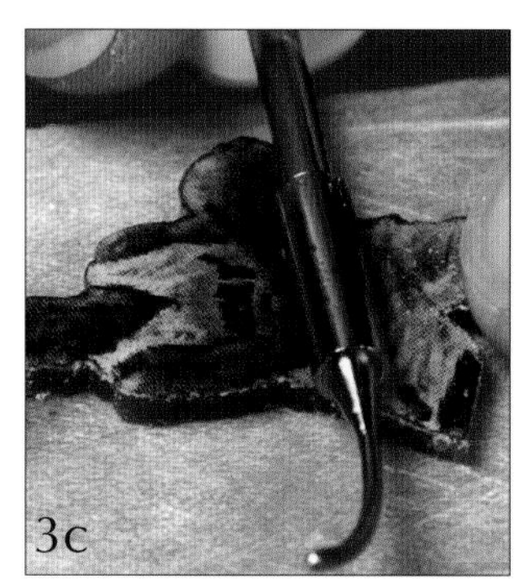
3c

## Determining the Model's Thickness

Many years ago I stumbled on a technique for measuring the thickness of a wax model in areas too small for a gauge. In instances such as these, I let the light show me how thick the piece is.

Begin by creating a reference gauge out of a piece of carving wax. Make a wedge that measures about 1 inch long and tapers to about 3 millimeters. Along one side of the wedge, cut in a series of steps, with each being about 0.5 millimeter thicker than the last.

When this gauge is viewed with backlighting, the steps show up as various shades (Figure 4). These shades can then be compared to the carved piece.

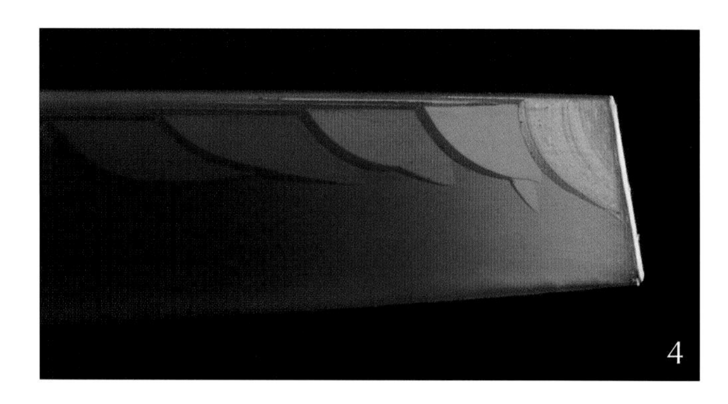
4

## Creating Mirrored Images

The easiest way I have ever found for creating a mirrored image is to make both sides as part of a whole and split them later. In this instance, the request was for a pair of puppy dog earrings. The first step required carving the puppy dog in 3-D and then sawing it in half (Figure 5a). After the pieces were separated I was left with both halves, one mirroring the other. The only steps remaining were to hollow out the backs of the pieces for weight (Figure 5b) and to add a sprue for casting.

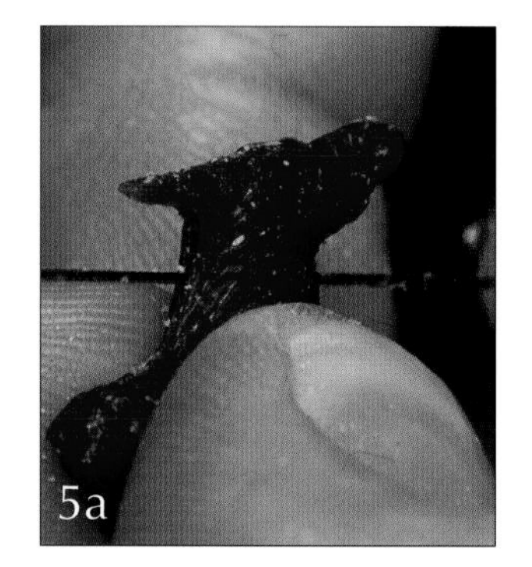
5a

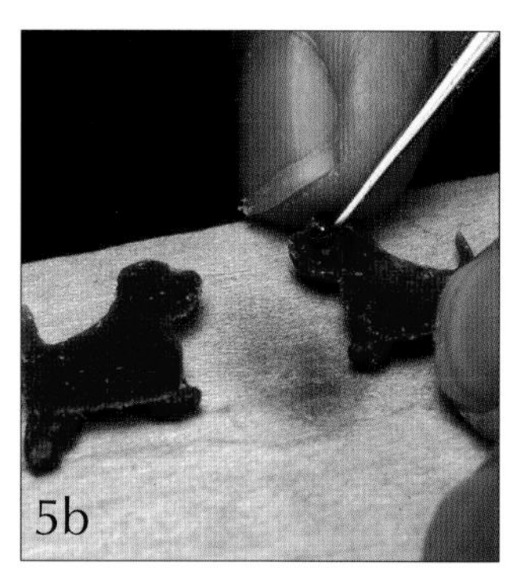
5b

# Heavy Metal

## CARVING WAXES FOR PLATINUM

BY MICHAEL BONDANZA

Suppose you awoke one morning to find that you were 40 percent heavier and packed into the same size frame. Your whole concept of who you were would be altered. Actions you took for granted—tying your shoes, walking, jogging—would all be affected by the increased weight. Just getting out of bed and dressing would be a major task. Talk about a nightmare!

Now project what might happen if you cast your favorite gold collection in all platinum. Platinum is dense, therefore heavy. It weighs about 40 percent more than 18k gold. You would need to change your whole concept of weight and cost of metal. Actions you take for granted would have to be reinvented—beginning with the model making process.

From a technical point of view, the solution to this weight problem lies in the design of the model. To cast platinum effectively, you need to make models precise and thin. This process, which I like to call "detailing," begins with a few simple questions. If you're planning to work in platinum, you should know the answers before even picking up your file.

***How can I determine the platinum weight of my carved wax?***

Weigh the wax and multiply by the metal's density (usually 21.5). If the wax weighs 0.05 dwt., for instance, then the casting will weigh 1.075 dwt. One-piece rings from 5 to 16 dwt., if properly sprued with heavy 6 gauge round wire in a "Y" configuration, often work well. I think 8 dwt. is the limit of consistent success, though; I might get 6 out of 10 good castings in production.

***How large a piece can I cast successfully?***

For large production runs where finished pieces comprise many small parts, such as a necklace, a 25 mm long by 14 mm wide wall with a thickness of 0.8 mm would be considered large. I prefer to use smaller parts—5 mm by 8 mm, for example, with a wall thickness down to 0.4 mm. I find that translates to 0.5 mm in the cast pieces due to the higher pressure of platinum casting. However, the normal 3 to 5 percent perimeter shrinkage still applies.

In general, I have had consistently good results with castings that measure up to 15 mm by 15 mm and have a wall thickness of 0.5 mm, using heavy 6 gauge round wire sprues. The upper limits of finished weight in any platinum casting is 8 dwt.; heavier pieces are possible, but don't expect to get good results with large quantities.

A note of caution: if you're making one-of-a-kind pieces from wax carvings that weigh in the range of 6 to 8 dwt. in finished platinum, your success rate may drop to 30 percent due to the unpredictability inherent in its uniqueness. To avoid needing to recarve your model, make an RTV silicone mold of it so that you can have several waxes from which to cast.

***What if a piece is larger than the recommended size?***

If the piece weighs 11 to 15 dwt. or higher in platinum, you might want to cast it as two or three sections. This will also give you an opportunity to clean and polish each section of the final cast piece before welding them back together.

***What else do I need to consider when making my models?***

Consider the choice of using either large parts that require less assembly labor or small parts that require more. I tend to go with smaller, lighter, cleaner castings and tailor my designs to accommodate this approach. Granted, casting techniques have become better, allowing larger parts with thin walls to become perfect finished pieces. But if you're accustomed to working in gold or silver, platinum castings will seem cruder, and cleanup will be much more demanding. Slight imperfections in the surface or shape will be amplified and more difficult to rectify. Even a fingerprint in the wax is a big deal—it's like a footprint in cement.

This goes to the heart of model making: each wax must be as perfect as possible, so that very little filing and cleaning will be needed on the finished piece. Always pre-drill and azure holes in wax as well as metal models, and apply a fine finish to all surfaces.

***It seems like labor-intensive work. Is there any alternative?***

Let's not kid ourselves: It *is* labor-intensive work. But if you approach design as a whole process and know platinum's characteristics, you'll be able to build anything you can dream of—and not wake up to a nightmare.

# Waxing Poetic

## CARVING A MODEL WITH HEART

BY CHUCK HUNNER

This whole thing started one day on the Florida coast. Seeing that the weather was perfect for running—bright sun, temperature in the low 80s, a gentle breeze carrying a fresh salt smell from the gulf—I decided to take a jog along the beach. After only about half a mile, I saw someone kneeling in the sand a few hundred yards ahead.

Kim, the turtle lady of Seagate beach, was cleaning out stragglers from a loggerhead sea turtle nest that had hatched three days earlier. I jogged closer, out of breath but excited. The little guys had not left the nest with the first wave of hatchlings, and Kim was giving them a second chance. She reached into the hole, up to her elbow, and lifted out each turtle.

That evening she released them under the protection of darkness. I took some photos of the event, getting front, back, side, top, and bottom views of the hatchlings. I had already decided I wanted to make a wax carving of this animal and preserve it in a pendant. A threatened species that had been on the endangered list until 1994, the loggerhead sea turtle has since made a comeback. Watching these hatchlings, I could see why: They had a lot of heart.

Just like our business. As jewelry manufacturers, we make things that allow people to express the abstract concept of love through gifts of precious materials. It seems natural to me to make things with a lot of heart that are going to be given with a lot of heart. And that process starts with the wax carving, my favorite part of the entire investment casting system.

I've been making jewelry on and off for 20 years, and during that time I've learned a few tips about wax carving. As jewelers, our craft (and science) involves a series of interdependent steps, and the final product clearly reflects how well those steps work together. By coordinating the wax modeling with the rest of the system, I've been able to achieve easier production, as well as a reduction in finishing time.

## The Model Audience

The first step to a successful piece is choosing your target market. The more focused your market, the easier it will be to design for it. The sea turtle pendant was aimed at an individual I know personally; she's an example of the affluent, caring people who live on and visit sea coasts. I know that if I could please her, many like her would also be pleased.

The choice of the baby turtle was driven by many considerations. It was a baby. It was a member of a threatened species, and the world community was focused on preserving wildlife. (The more the species' image could be shown, the more interest the public would have in protecting it.) And nautical ideas sold in Florida, especially in the winter. The state spent plenty of money keeping its citizens informed about sea turtle nesting habits. Everywhere near the beach, bumper stickers and signs displayed pictures of baby sea turtles. This was free advertising!

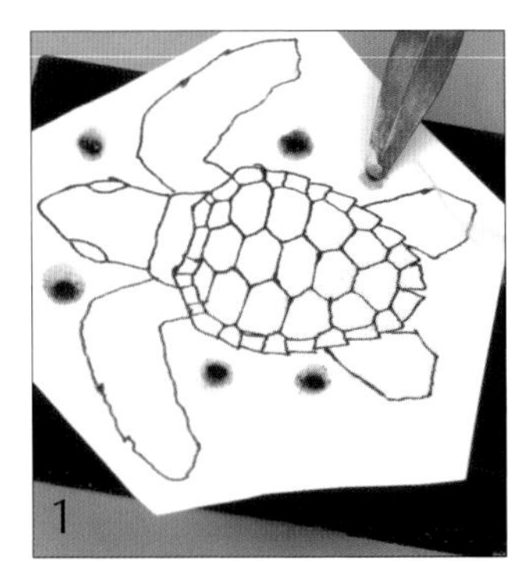


In addition, the sea turtle design made sense from a manufacturing standpoint. It's easier to cast smaller expanses and textured areas than it is to cast a large, flat, shiny surface without any porosity. (Minor porosity can be easily hidden by textures.) The turtle, though a slightly labor-intensive piece with all the needed soldering, allowed me to keep the labor to a minimum and use the tumblers to their best advantage.

When considering your target market and design, you also have to take into account the size of the piece (weight and thus price). I wanted to take the 3-D approach and carve the turtle about three quarters life-size. If the piece was solid (about 40 dwt. in 14k) it would be a very heavy pendant, and too expensive for most of my customers. That meant the charm needed to be hollow. However, if I hollowed out the back like so many other charms on the market, I would lose some uniqueness. I've noticed that most customers will turn over a charm, and they're always delighted when the back is fully finished. So I incorporated this into my design as well.

I then planned how I wanted the pieces of the charm to go together. While some manufacturers just get an idea and start carving, I've always done my best work by making sketches first. It's like studying a map before going into an unfamiliar neighborhood: You know the address you're looking for, and if you trace your route before you start, you'll avoid traveling

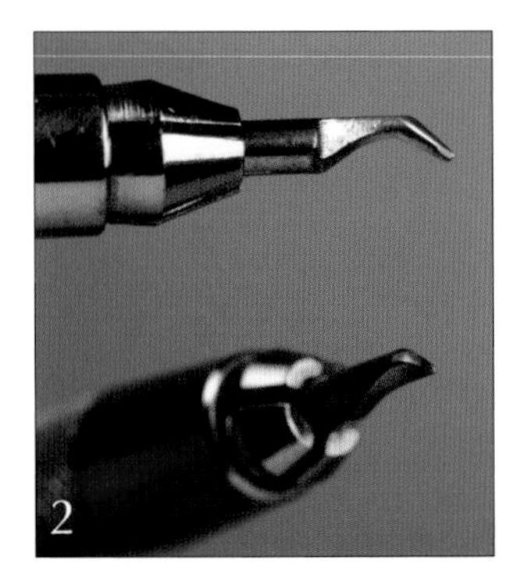


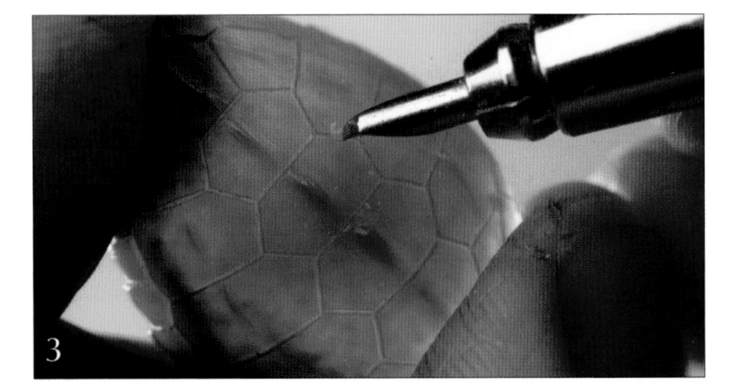

deadend streets. The sketch may be a drawing or just a tracing from a photo, sized up or down with a copy machine.

The sea turtle design, I decided, lent itself to a two-toned approach, with the shell one color and the limbs and head another. A combination of gold (yellow or green) and white (either white gold or platinum) not only would be trendy, but also would accentuate the shape of the turtle hatchling. The design would have some volume but not be overly round, and from a distance its teardrop shape would appear interesting. I sketched my ideas, and moving my pencil along the contour of the turtle helped me learn the volume of the piece. This learning bears fruit during the carving, since much of the model making process has to do with perceiving the volume that is being defined.

It's important to know how the volume of the model will interact with the fluids that fill the cavity. When I'm carving wax (making a positive space), I'm also creating the mold or investment cavity (making a negative space): When I cut a groove in the wax, I also see a narrow ridge being created. What happens during the carving of the wax influences the speed at which molten wax or metal will flow through the volume. Thick areas with high volume-to-surface-area ratios can take a long time to solidify. This could cause dimples in the wax, or dendritic porosity in the metal. Thin areas with low volume-to-surface-area ratios can solidify more quickly, which could result in cold shunts or incomplete fills.

The ideal cast shape looks like a wedge with the gate at the wide end. This allows the molten metal to fill the form and progressively solidify from the thin edge of the wedge back to the gate and into the sprue. I do my best to imagine the model as one of those wedges, so that the areas farthest from the gate are thinnest and the areas nearest the gate, thickest.

## The Kindest Cut

Once my sketch was done, I transferred it to a polished block of hard wax, making sure to keep the paper steady (Figure 1). I pushed an incandescent wax pen tip through the paper and into the wax at several spots around the design. The wax melted onto the paper and, when cool, held the paper securely while I traced over the design. (I like using hard green carving wax because it holds

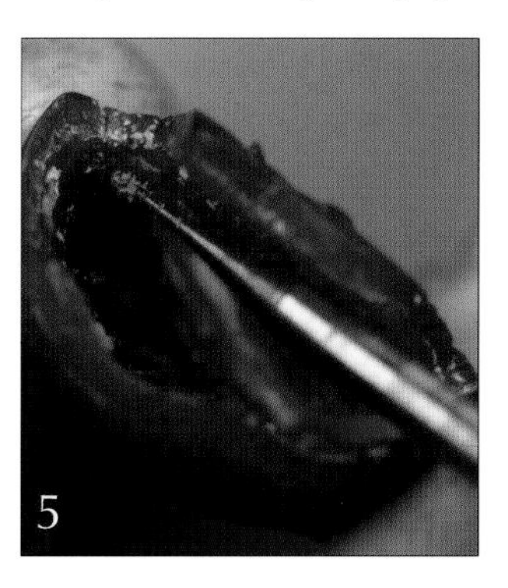
5

details better than the softer, more flexible waxes. It's also easy to file and cut to shape, since this brittle wax won't clog the tools.)

For carving, sharp tools that cut the wax rather than mush it provide the key to precision. As I did with the turtle, you should use rotary files and wax burrs to do the rough shaping, and round, flat, and onglette gravers to cut details cleanly. You can also make great custom knives from broken or dull burr shafts. To make such a knife, grind and polish the end of the burr shank to the shape of a hockey stick (Figure 2). The resulting tool will have three sharp edges: a long outside edge, a short tip edge, and a mid-size inside edge. The outside edge shaves wax as the knife is pushed away from the carver. The tip scrapes channels or slices very fine detail. The inside edge cuts toward the carver. Tailor the angle of the cutting face to that at which you most comfortably hold a pen. Clamp the finished knife in a draftsman's lead holder.

You can detail carvings all day with a tool that's easy to hold. But remember: Sharpness is important. A sharp blade leaves a shiny slice and will chatter when scraped across the wax. A slightly dull, polished blade will leave a shiny scrape. If a knife is highly sharpened and then slightly dulled with a hard felt wheel and tripoli, it will leave a shiny cut no matter how it's used.

To incise detail, I used a round-nosed graver (Figure 3). This easy-to-control tool made the bottom of the inscribed lines round, which would prevent investment from being trapped in a V-shaped channel. Also, the graver shaved the wax rather than scratching or plowing it, and the resulting line had much cleaner edges.

Once I finished detailing the individual waxes—legs, head, mouth, shell—I next had to separate the shell into two halves. To do this, I used a curved pin tool sharpened into a square, like a reamer (Figure 4). Once I had shaved deeply enough (2 to 3 mm) into the piece, I began to burr from the belly out toward the incision (Figure 5). I used a smallish wax burr (about 1 to 2 mm in diameter) to hollow out the turtle design. Once the inside edge of the hollow extended all the way around to the incision, the halves separated easily.

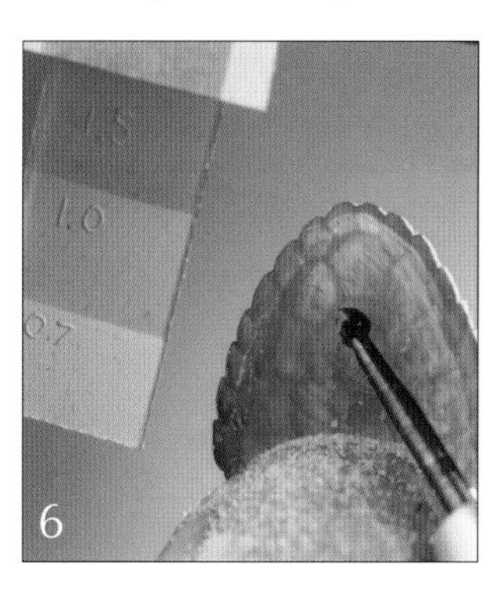
6

Another way to cut a charm in half is with a 0.5-inch length of 3/0 saw blade held in a pin vise. This width blade makes a thinner, more precise cut than the scribe. However, the procedure will work only if you have room on the inside of the charm for some "throw" in which to move the saw blade back and forth. This trick also works well for cutting apart nested wedding ring sets while they're still in the wax, as well as for doing wax filigree work.

Once the shell halves were separated, I finished hollowing. To determine when the carving had been cut to the desired thickness, I held it in front of a light source and used the wax's color as a guide. (Thick walls prevent light from shining through, thin walls permit it.) I filed a 1 inch by 0.5 inch piece of the carving wax to the thickness I wanted, then taped this wax tile to the edge of my bench light and used it as a comparison to the color of the carving (Figure 6).

I thinned the walls of each half to a thickness between 0.8 and 1 mm, cutting on an angle so that the inner half fit into the outer half like a plug. I compensated for the little bit of wax lost in the cutting process by making the inner half a little thicker. If the work is done precisely, the two wax halves will fit well.

I finished the surface of the wax first with 400, then with 600 grit emery paper. Fine nylon, such as that found in hosiery, is also a good prefinisher. And if you don't need precise detail or aren't concerned with a very high polish, you can heat polish the wax. I'd normally take a shortcut and heat polish the model with a tiny flame or an incandescent wax pen tip, which would melt the wax's surface. However, this technique may cause warping when you thin the wax down to the final model thickness. (With my turtle, I heat polished the head and limbs since it didn't matter if they warped slightly.)

For the final polish, I rubbed the carving with blotter paper. Stiff blotter paper keeps the corners of incised designs square, as will thin cardboard soaked in wax solvent. If you use the wax solvent with a soft surface, like a cloth or a paper towel backed by your finger, you will round off the corners and make them less distinct.

This problem also occurs later in the manufacturing process: How do you keep the details sharp, yet finish the surface of the design to a high gloss? The answer can be found in your design. Since the tumbler is going to wear away the edges more aggressively than the flats, exaggerate the edges.

When I was done polishing, I lightly buffed the final wax surface to a high sheen with the soft side of a paper towel (without the solvent). By following these steps, I ended up with a good wax model—and a good wax model will usually lead to a good finished piece, with all parts of the process working together harmoniously. You might even say that a good model is the heart of the system. And that's what our business is all about: heart.

TRADE TIPS FOR

# Working with Molds

COMPILED BY SUZANNE WADE

## Avoiding Bad Molds

You thought you did everything right: You tightly packed the mold frame with rubber, positioned the master model within it just so, vulcanized the rubber at the proper temperature—and still you have a bad mold.

And if the mold isn't good, there's no way the cast piece can be good. So what happened? To help you figure this out—and keep it from happening again—Michael Knight of Castaldo in Franklin, Massachusetts, offers this troubleshooting checklist:

**Finished mold is tacky and soft.** The cause is likely insufficient vulcanization time or temperature. Be sure to observe the time and temperature recommended for the rubber by the manufacturer. Check the calibration of the vulcanizer's temperature settings with an accurate thermometer, since vulcanizer dials can be wildly inaccurate.

**Finished mold is hard and springy, so it won't lie flat.** The cause may be excessive pressure during vulcanization or excessive vulcanization time and/or temperature. Reduce vulcanizer pressure, observe the time and temperature recommended by the manufacturer, and check the calibration of the vulcanizer's temperature settings.

**Partial de-lamination of mold into separate layers.** The most likely cause is under-packing. Another possible cause is contamination with hand oils, silicone spray, talc, etc. Unfortunately, you won't be able to salvage this mold.

**Air bubbles throughout and/or large depressions in top and bottom surfaces.** Your mold frame is underpacked. Pack the mold frame more fully in the future. Do not rely on counting the number of layers of rubber to ensure a fully packed mold frame, since the thickness of the layers can vary.

**White powder on unvulcanized rubber.** This isn't actually a problem; it's perfectly normal. You can disregard the powder.

**The rubber is hard and won't vulcanize.** Your problem is probably that the rubber had already been fully or partially vulcanized through accidental exposure to heat, or through aging of the material. Toss it out and order some new rubber.

**The rubber is hard and stiff.** The rubber may feel warm, but it may have frozen through long exposure to cold. It can be restored by warming slowly at approximately 100°F (38°C).

**Excessive shrinkage.** There are many possible reasons. For example, the vulcanization temperature could be too high. Be sure to observe the time and temperature recommended for the rubber by the manufacturer, and check the calibration of the vulcanizer's temperature settings. If shrinkage is still a problem, you can also try reducing the temperature to 290°F (143°C) and doubling the time.

**Rubber does not flow into all cavities.** There are two main causes for this problem: The mold is not packed properly or the vulcanizing temperature is too high. Proper packing generates the internal pressure in the mold frame needed to drive the rubber into full contact with the model, so an inadequately packed mold frame may not push the rubber into all cavities. Packing all cavities with scraps of rubber should alleviate this problem.

If the vulcanizing temperature is too high, the rubber will cure before it flows into all cavities. The rubber manufacturer's recommended temperature and time are designed to keep the rubber in a liquid state long enough to fill all cavities and still be fully cured at the end of the vulcanizing cycle. Observe the time and temperature recommended for the rubber by the manufacturer, and check the calibration of the vulcanizer's temperature settings.

## Getting to the Root of the Problem

Talc has long been used for powdering molds, but it can cause health problems when breathed in. Good ventilation can cut down on the danger, but fans can do only so much. And since talc is a mineral, it won't burn out in a kiln if it gets on the wax, potentially leading to surface problems on your castings.

With so many strikes against it, John Henkel of J.A. Henkel in Brunswick, Maine, decided he just didn't want to use talc in his shop. He tried a couple of alternatives, including cornstarch and flour, with limited success. Then he found arrowroot powder in a health food store. John writes:

"Arrowroot seems to be the right size and it doesn't gum up. I sparingly tap the powder around the back side of the mold and on the piece itself. Before returning the mold to storage, I use compressed air to dust out the powder."

# Fitting the Mold

## COMPARING NATURAL RUBBER AND SILICONE

BY SUZANNE WADE

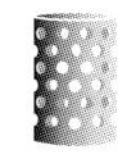

Natural rubber or silicone rubber? In the world of lost-wax casting, this question is hardly an academic one. The mold material chosen must be up to the task of turning a single original model into hundreds or even thousands of wax models for casting, and choosing the wrong material can lead to some serious hair-pulling in the casting department.

There are actually three primary categories of rubber mold making materials: natural rubber, vulcanizing silicone rubbers, and room temperature vulcanizing (RTV) rubbers, a non-shrinking, no-heat option. Each has its advantages and disadvantages, and as a result, each may be the best choice in certain situations. The key to choosing between them is to understand the properties of all three, and then decide which to use based on what qualities are most important for a given situation.

## Au Naturel

Natural rubber's advantages include high tear and tensile strength and a long shelf life. As the industry standard for decades, it's also a more familiar material to many mold makers.

Tensile strength is commonly measured by the number of pounds per square inch (psi) required to tear a standard sample. Natural rubber can withstand up to 3,000 to 3,500 psi, while silicone rubbers can handle 1,000 to 1,400 psi. RTVs are the most prone to tearing at 100 to 200 psi.

This greater tensile strength means that natural rubber molds normally last longer in production situations. While natural rubber molds may be used thousands of times before deteriorating, silicone molds typically withstand hundreds of uses, while some putty-type RTVs may be usable for only 10 to 30 waxes.

Because they have been in use for decades, natural rubber molds have been proven to remain usable for up to 50 years under ideal storage conditions. Although silicones may demonstrate similar staying power, their more recent introduction (in the 1970s) means their storability for long periods of time is relatively untested. RTVs can be less stable, because some are susceptible to moisture and will deteriorate more quickly if exposed to atmospheric humidity. (Pre-vulcanized shelf life for natural rubber and vulcanizing silicone rubbers is about the same at one year. RTV silicones are typically good for approximately six months.)

Although natural rubber was initially less expensive than silicones, the price of vulcanizing silicone rubber has dropped and the two are now nearly equal. RTVs, however, remain more expensive than either natural rubber or vulcanizing silicones.

## The Silicone Story

Since its introduction to the jewelry industry, silicone has been making steady inroads into the mold maker's shop. The material's advantages include a higher-quality surface finish and easier release of wax models, as well as the availability of RTVs.

Because silicone rubber molds are self-lubricating, they don't require the use of spray release agents, reducing problems caused by the buildup of these agents on the molds. This property also makes it easier to release intricate, detailed designs from silicone rubber molds on the first try, versus the tinkering occasionally required to get the right amount of release agent in a natural rubber mold.

In addition, silicone rubber molds typically produce a shinier wax model, which can result in a higher-quality surface finish on cast items. And because silicone is relatively inert compared to natural rubber, it will not react with silver or with the copper in sterling, reducing the need for nickel or rhodium plating of the models. (Certain materials, such as some plastics, will inhibit curing of RTVs. In these cases, coating the model will frequently solve the problem.)

Some users find that silicone rubber holds its shape during wax injection better than natural rubber, making silicone more tolerant of variations in injection pressure. Also, silicone rubbers typically produce less flashing, since seals can be tighter.

Silicone's putty-like texture makes it easier to pack a mold, since there's no need to cut the rubber to size, and it is also easier to cut. In addition, different hardnesses of silicone rubber can be combined in a single mold and vulcanized together.

Silicone rubber's higher heat resistance makes it suitable for casting metals with melting points of less than 315°F (157°C), such as pewter and tin, directly into the mold. Silicone rubber also has a higher vulcanizing temperature than natural rubber—330°F to 350°F (165°C to 177°C ), compared to 310°F (154°C) for natural rubber.

## FYIs about RTVs

A third option for mold making is the non-shrinking, no-heat RTV rubber. This two-part mold material is typically mixed and poured around the model, then allowed to cure for 18 to 72 hours.

RTVs offer both advantages and disadvantages over traditional natural rubber and vulcanizing silicone molds. Room-temperature curing permits the molding of fragile materials that would be damaged by the pressure of a vulcanizer, such as hollow beads. RTVs also typically offer 0 percent shrinkage, compared to 0 to 4 percent for natural rubber and 2.6 to 3.6 percent for vulcanizing silicones. (This is a particularly crucial factor when the manufacturing process involves stone setting.)

Clear liquid RTV silicones also provide greater ease for inexperienced mold cutters, since the mold maker can see the model through the silicone rubber.

On the downside, RTVs usually have significantly longer curing times—hours or even days, compared to 45 minutes or less for natural rubber and vulcanizing silicone molds. They also offer the lowest tensile strengths of all the common mold materials, and require careful cutting and gentle use to avoid damaging the mold.

## COMPARING RUBBER MOLD MATERIALS

| MOLD MATERIAL | VULCANIZING TEMPERATURE | CURING TIMES | TENSILE STRENGTH | SHRINKAGE % |
|---|---|---|---|---|
| Natural Rubber | 310°F (154°C) | 45 minutes or less | 3,000 to 3,500 psi | 0 to 4 |
| Vulcanizing Silicone Rubber | 330°F (165°C) to 350°F (177°C) | 45 minutes or less | 1,000 to 1,400 psi | 2.6 to 3.6 |
| RTV Silicone Rubber | N.A. | 18 to 72 hours | 100 to 200 psi | 0 |

Many RTVs must be mixed in precise amounts, and the working time for the molds is often quite short at just a minute or two, although there are some RTVs that offer work times of up to 60 minutes. Most liquid RTVs require vacuuming to remove air bubbles, as well.

Whatever your choice for mold making, you'll need to use the right tools and techniques to get the best possible results. When cutting molds, be sure to use a sharp blade, and replace it immediately if it gets nicked or dull. Make sure that your hands are clean, and that the model has no visible defects.

If you are using a vulcanizer, double-check the accuracy of the vulcanizer's temperature gauge with a thermometer. Check each plate separately by placing a block of scrap wood into the vulcanizer, and sandwiching the thermometer first between the wood and the top plate, and then between the wood and the bottom plate. Observe the temperature variation through the vulcanizer's entire heating and cooling cycle.

And whatever mold material you use, be sure to follow the manufacturer's instructions precisely. While some materials are more forgiving than others for variations in process, all will offer better consistency and predictable results if you follow the steps the manufacturer has outlined for it.

So what type of rubber should you choose? The question doesn't have a single answer, and with technology improving all the time, the answer you come to today may not be the best one for tomorrow. Natural or silicone? The choice is yours.

*This chapter was adapted from an article that appeared in the May 1998* AJM. *Those industry experts who contributed their knowledge to it included Dominic Annetta, Steven Blythe, John Davidian, Elaine Corwin, Roger Greene, Michael Knight, Jeffrey Mathews, Lee Mosemiller, Bob Romanoff, Liz Rutherford, and Len Weiss.*

# Moving to Metal Molds

## EVALUATING THE BENEFITS AND THE PITFALLS

BY SUZANNE WADE

In an age where productivity and consistency can determine whether a jewelry piece is profitable or not, many manufacturers are discovering a technique first explored during the rubber shortages of World War II—the use of metal instead of rubber to create molds.

Metal molds are particularly useful in applications where tight tolerances are needed, such as lightweight and promotional lines that must hit a price point. Because rubber molds expand with use, pieces made with these molds can gradually become heavier over time. Metal molds offer a hard surface, so they can be injected at higher pressures without fear of gaining weight: Metal molds can withstand pressures as high as 3,000 to 6,000 psi, compared to the 30 to 40 psi common among most wax injectors.

Such high pressure also helps pieces fill better, which permits the creation of designs with both very fine detail and thin walls, giving a big look at a lower price. The high pressure overcomes friction, allowing even very thin pieces to fill. The same pressure on a rubber mold would cause compression of the rubber, causing the part to get bigger and distort.

The consistency offered by metal molds can also improve productivity. Where rubber molds may overheat or otherwise develop problems that prevent good wax injection, metal molds will consistently pump out identical waxes, without ever showing the kind of temperamental nature that can make working with rubber so challenging. This consistency often allows good operators to increase their production from 500 pieces a day to as many as 2,500 pieces a day.

Finer detail, better consistency, lower price points, higher productivity...all fine reasons for replacing rubber and moving to metal. However, before switching any part of their casting production, manufacturers must also be aware of certain drawbacks. For example, not every piece of jewelry can be made with a metal mold, and even those that can will often raise labor costs. As with any change, the proof lies in the details—and metal molds have many.

## Soft-Pressed vs. Hard Metal

There are two primary types of metal molds: soft-pressed molds and molds milled from harder metals, such as brass, bronze, aluminum, and steel. Soft-pressed metal molds are made from bismuth-based alloys with melting points of 115°F to 500°F (46°C to 260°C), which are poured over and around a model and pressed in a vulcanizer to create an injectable mold. In hard metal molds, the mold cavity is cut directly into the metal by means of a computer-driven milling machine.

CAD/CAM technology has made the creation of hard metal molds much simpler than in the past. In former years, manufacturers would often make a hard metal mold by hand-tracing etchings on acetate using a compass, protractor, and French curve. The acetate tracing was then used as a pattern with a pantograph to machine the molds.

This early approach was not only low-tech, but also time-consuming and labor-intensive. Today, CAD/CAM allows tool makers to start with a scanned two-dimensional drawing from the design department, which is converted to a 3-D pattern in the CAD program. The CAM software then takes the 3-D model and converts it to toolpath code for the milling machine, which will cut the negative cavity in metal.

Thanks to the precision of computers, CAD/CAM systems produce precise mold pieces that fit together without so much as a visible parting line. In addition, the same files can be used to cut carbon fixtures for oven soldering that will precisely fit the mold's product, speeding final assembly.

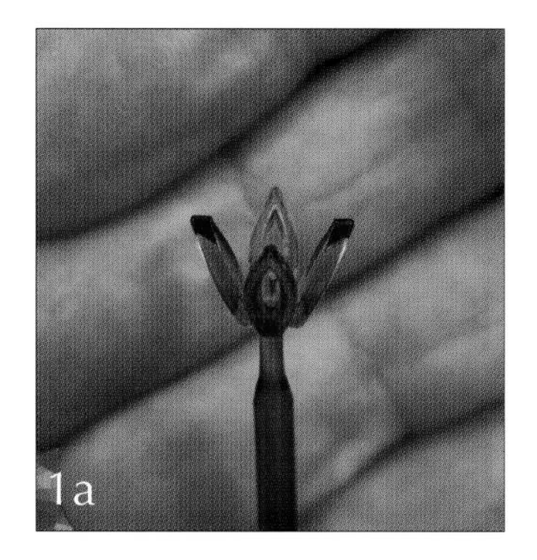
1a

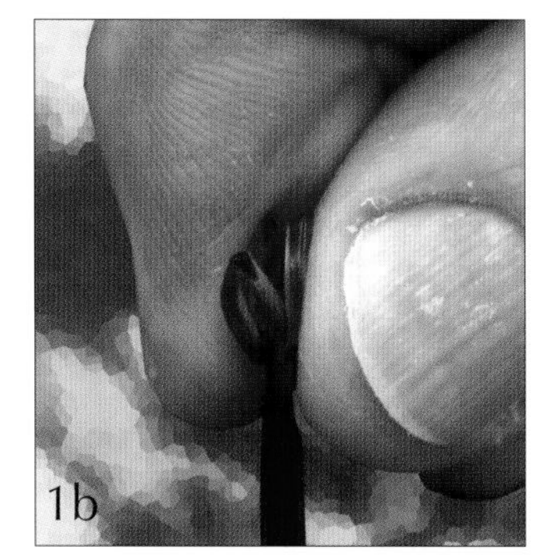
1b

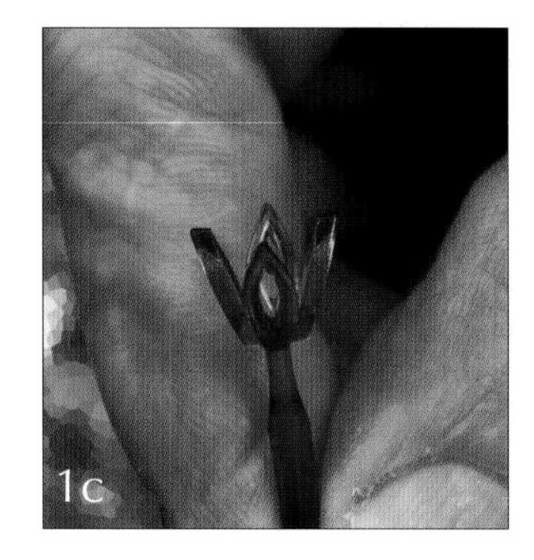
1c

Not all CAD/CAM systems will work for milling hard metal molds, however. In making metal molds, the parts must fit together precisely. The milling machine must be able to produce the pieces of the molds within very tight tolerances—which normally means a high quality milling machine.

And quality comes with a hefty price tag. It can cost $30,000 to $40,000 to buy metal mold making equipment, including up to $8,000 just for the plastic injection equipment required by metal molds. Compare that to the $500 price tag of equipment for rubber molds, and the benefits of hard metal molds take on a new perspective.

For manufacturers who want a less expensive alternative, soft-pressed metal molds may be the answer. Their manufacture does require some special equipment, such as a custom-machined case to hold the metal as it is pressed in a vulcanizer, and a temperature-controlled melting pot. It also requires a real rather than a virtual model for shaping the mold, and that model must be carefully finished for best results. However, there is no need for expensive CAD/CAM equipment.

The downside to soft-pressed molds is that they are more fragile than those made of hard metal, and they require careful handling. They can easily be damaged if dropped, and a single ding in the surface can ruin the mold.

Estimates of the number of pieces that can be produced before the molds are refurbished range from 500 to 5,000, with some users reporting up to 50,000 pieces from a single soft-metal mold. The variation is due in part to the fragile nature of the molds. Because metal molds are often made in multiple sections that lock together, even a tiny amount of wear or damage will throw everything off, requiring the manufacturer to rework the entire mold.

Deciding whether to use soft-pressed or hard metal is just one of the choices manufacturers must make. In addition, they must choose which material will be injected into the mold: wax or plastic. Both have their advantages and disadvantages.

Wax is simpler, faster to clean up, and, for many manufacturers already accustomed to it, easier to work with. Plastic requires aggressive solvents for cleanup, but it's more pliable and stores better. It's springy, and will return to shape after it's squeezed or bent instead of breaking, as wax will (Figures 1a-1c). Plastic can precisely replicate details without distortion. In addition, plastic cools faster than wax, and can be handled much sooner. Unlike wax, plastic doesn't become brittle or tear easily.

Despite these differing characteristics, production procedures are nearly the same for both wax and plastic. The primary difference is that wax injectors permit the operator to either hold the mold manually or use a clamp, while plastic injectors generally require a clamp.

Plastic can also be invested and burnt out using normal casting procedures, with the exception of steam dewaxing, which isn't effective on plastic. However, an investment with a higher proportion of calcium sulfate is often needed to prevent such problems as spalling, even though this investment makes breakout more difficult.

## The Minuses of Metal

While metal molds may offer many advantages, they do have their limitations—and companies must consider them before making any final decisions. Not every manufacturer sees big productivity gains, for example. In fact, some companies have found their production actually dropped with the introduction of metal molds. While rubber molds require the same amount of time to shoot a wax no matter how complex the piece, metal molds for such pieces are normally created in multiple parts, which must be assembled prior to wax or plastic injection. This assembly time may actually reduce the number of models the wax department can produce.

Metal molds also take longer to create, and as a result cost more. While a rubber mold normally takes approximately an hour and a half to make, a complicated metal mold can take a couple of days. Manufacturers must weigh carefully whether the additional time and expense is worth it: For some, the cost of the mold will outweigh any productivity gains. For others, the lower gold loss attributable to the consistency of metal molds will more than make up for the additional time and expense.

Perhaps the most important consideration is that not every piece of jewelry can be reproduced in a metal mold. Undercuts, for example, pose special problems. While rubber or silicone will flex to release an undercut piece, a metal mold doesn't yield, making such a piece difficult if not impossible to remove. This problem can be avoided by creating a mold with more sections, enabling it to come apart and release the pattern more easily. However, the greater the number of undercuts, the more sections that will be needed.

Although in theory almost any piece of jewelry can be made if the mold has enough parts, metal molds are most useful for symmetrical pieces, such as pendants or charms. The simplest design may need only a two-piece mold, with two halves that split the product, while more complex shapes might require the mold to split in up to eight directions. As a jewelry design becomes increasingly complex, the mold to make it can begin to resemble a Rubik's cube—and be nearly as challenging to assemble.

It's also important to consider whether the waxes you create in the metal mold can be successfully cast. For example, although a metal mold will permit the creation of extremely thin patterns, these pieces can create headaches in the casting department. Such thin pieces may also be easily damaged during subsequent operations.

Given the variables, then, should companies move at least some of their production into metal molds? That's a decision that can be made only by individual manufacturers. Switching to metal molds can improve productivity when your designs are simple and symmetrical, but may actually slow production if the pieces are very complex and require multi-part molds. Only careful examination of a particular product line, and consideration of the company's production volume and needs, can determine whether metal molds are a potential nightmare or the stuff dreams are made of.

*This chapter was adapted from an article that appeared in the August 2000* AJM. *Those industry experts who contributed their knowledge to it included Gary Ayvazian, Alex Benedict, Greg Gilman, Timo Santala, Jonathan Seidel, and Chuck Wolfmueller.*

# Finding Flaws

## IDENTIFYING AND PREVENTING WAX DEFECTS

BY J. TYLER TEAGUE

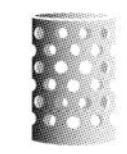

In any kind of manufacturing, there are defects in materials, processes, or products that need to be corrected. Ideally, problems should be prevented early in the process, rather than fixed at the end—especially since the final rework stage is one of the most costly labor steps in jewelry manufacturing.

Of course, to prevent them, you must know what causes them. And that's not quite as simple as it sounds—for casting, in particular. Manufacturers often waste thousands, perhaps millions, of dollars trying to diagnose defects in the casting process, when actually the flaws can be traced back to one of the most overlooked areas: the wax department.

What follows are some of the most common wax defects. Once identified correctly, these problems are usually simple to fix, saving you money immediately.

## Air Bubbles in the Wax

Bubbles under the surface of the wax appear either when air is injected into the mold along with the wax or when air in the mold is trapped during the wax injection process. In either case, an air bubble can pop when the casting investment is vacuumed, and when it does the void fills with the investment slurry. Later, the metal fills around these formations and—*presto!*—investment inclusions, which most often appear as small, rounded holes in the castings.

You can inspect waxes for the presence of bubbles by holding them up to a small table lamp. The bubbles will appear as light spots, although how easy it is to see them will depend on the "readability" of your wax. You can then pop them with a heated pointed tool and fill the exposed holes with wax. Due to the flow dynamics of wax in a mold, these defects will tend to show up in approximately the same places on all your waxes or castings of the same style.

A better solution is to eliminate the source of the bubbles whenever possible. How that is accomplished depends on whether the air is being trapped during the wax injection process or is being introduced with the wax.

**Trapped air.** If air doesn't escape the mold at an equal or faster rate than that at which the wax is being injected, it will become trapped. These types of trapped air bubbles can be reduced or prevented by the proper application of powder in mold vents, which allows air to escape from the mold cavity during injection. Reducing the injection pressure and slowing the injection rate will also help.

Air can also be trapped in molds if, because of the way the model is gated, the wax must flow backward turbulently to fill the pattern. This type of gating configuration can cause problems in casting as well.

**Introduced air.** Several sources can introduce air into the mold. One source is dissolved gas in the wax. Another is air that becomes trapped when new wax chips or dots are loaded into the wax pot: If small spaces between the cold chips are covered by a molten surface before the air can escape, these bubbles can be injected into the mold.

To overcome these challenges, pre-melt the wax in a reservoir unit and vacuum de-gas it before loading it into production wax pots. To do this, you will need a transparent lid with an airtight seal and a fitting for a vacuum line, as well as a separate vacuum release valve. If you don't have these resources, you could vacuum de-gas the production pots; just remember to isolate the pressure gauge on a wax pot during the process. (To do so, install a small ball valve between the gauge and the wax pot, and close the valve before starting the vacuum pump.) Also, never de-gas a full wax pot, or you will likely end up with wax in the vacuum hose.

Another source of introduced air is water vapor from condensation in compressed air lines. The pistons of an air compressor heat up during use, warming the air. This warm, humid air travels along the air pipes into the often air-conditioned wax injection room. The pipes are cooler in this room, so water vapor condenses and collects. This water eventually makes its way into the wax pot, where it will provide hours of air-bubble-repair enjoyment. To avoid this problem, you need to dry the air going to your wax pots. The use of line dryers, desiccant dryers, or a combination of the two is highly recommended.

Under certain conditions, automatic vacuum wax injectors can also cause bubbles in waxes. This is a source that most manufacturing companies don't consider and wax pot manufacturers don't want to discuss. Tool suppliers might tell you that a new automatic vacuum wax injector will cure your wax injection problems and increase your production. It would be more accurate to say that a well-engineered, properly aligned and maintained, auto-clamp vacuum wax injector can help produce waxes of consistent high quality and weight if used with properly prepared molds.

Waxes produced by automatic vacuum wax injectors are generally more consistent than those shot by hand on conventional wax pots. That's because the parameters are easier to repeat: Once you establish the proper combinations of wax injection pressure, clamp pressure, and forward force against the nozzle for each mold or mold group, automatic wax injectors can increase your production through decreased rejection rates.

But much depends on the mold itself. Most of these machines are designed to remove air from the mold by drawing a vacuum through the injection nozzle. However, if you are shooting a mold designed for traditional wax injection pots—one that is vented and powdered, for example—the vacuum injector simply draws in air through the vents. (The same problem occurs with

1

unvented molds that do not make a complete seal on the injection nozzle.) If air gets into a mold through either a vent or a leak, you gain nothing from the vacuum stage of the injector's process. Rather, some air will be drawn through the molten wax in the injection nozzle, creating small bubbles that will then be injected into the mold with no chance of escaping.

## Improper Pressures

An inadequate escape route for air during wax injection can also cause the wax pattern not to fill completely. When this occurs, workers are often tempted to adjust the injection, clamp, or forward pressures, or all three. However, this may only contribute to the wax bubble problem.

For example: The pattern doesn't fill, so the worker turns up the injection pressure. The injection pressure is now so high that the worker must apply greater clamping pressure on the mold to keep the wax from leaking out the sides. This additional clamping can close off small vents, further trapping air in the mold.

The worker may also raise the forward pressure. But too much forward pressure will cause the mold to split at the injection nozzle, allowing air into the wax stream. Because of this tendency, additional clamp pressure must be applied to keep the wax from leaking—which, again, will lead to small vents being sealed and air being trapped.

These kinds of defects are more commonly associated with hand wax injection because of its variable nature. This is where auto-clamp systems are a big plus; they allow you to discover, document, and repeat the ideal pressure combinations, eliminating the three pressures from the list of suspected defect sources relatively easily.

## Powder & Spray Procedures

Part of the wax department training that is often incomplete is the proper use and application of powder and silicone spray for rubber molds. Silicone sprays act as lubricants or mold release agents to facilitate easy, distortion-free removal of the wax from the rubber mold cavity. Powder creates a microscopic path for the air in the mold to escape through the vents as wax is injected. Each of these products must be used correctly for best results.

Ideally, the powder used should be very fine and organic (burnable), such as cornstarch or rice flour. Talcum powder is commonly used in the wax injection areas of the jewelry business because it is not as susceptible to humidity as cornstarch. However, talc is a mineral and does not burn out if it gets on or in the wax. Because it is lighter than metal, it floats to the surface of the molten metal stream and causes a grainy, rough surface on the casting (Figure 1).

Worse yet, many factories allow talcum powder to be applied with a "powder hammer," a piece of cloth wrapped around the talcum powder and tightly bound with string or rubber bands. Not only does this "tool" indiscriminately apply powder throughout the mold cavity, it also applies whatever dirt or foreign matter that happened to be on the table where it was laid.

Another potential problem is the misuse of silicone mold release sprays. I have been to factories where the liberal application of powder is immediately followed by a good dose of silicone spray, or vice versa. The powder traps or absorbs the liquid of the silicone spray, making a positive formation in the mold cavity. When the mold is injected, the wax flows over and around these formations of wet powder. This results in a pitted surface that, in extreme cases, can have an appearance similar to that caused by investment erosion or spalling (Figure 2).

These wet powder formations usually remain in the mold when the wax is extracted, and they duplicate the same surface defect over and over if not removed. Even if the clump of wet powder stayed with the wax when it was pulled from the mold, the powder would likely be washed out during the investing process and show up as a negative space in the casting surface.

The overuse of silicone mold release spray alone can also produce surface defects. These defects, which appear as smooth, rounded, irregularly shaped negative spaces in the casting surface, are caused when wax is injected into the mold cavity before the mold release spray has had time to dry. The wax pushes the droplets around before ultimately forming around them. I have seen this defect incorrectly identified as gas porosity several times.

To reduce the occurrence of such problems, make sure rubber molds are handled correctly. First, switch to cornstarch or rice flour, which will burn out cleanly if it accidentally gets on the wax. [Editor's Note: See also "Getting to the Root of the Problem," page 23.] To overcome the humidity issue, keep the powder in a sealed container when not in use, and keep a small porous packet of desiccant in the container. Just don't forget to replace or regenerate the desiccant packet from time to time.

When applying powder, apply it only to areas where the wax and the powder will have minimal contact. While bending the mold so that the vents are open, apply a light dusting of powder into the vent areas using a small, soft brush, then use a compressed air gun with "dry" air to blow out the excess powder. Release the tension on the mold so that the vents close, then spray with a very fine, light mist of a silicone mold release agent. Avoid sprays that spit larger droplets onto your molds.

Cleanliness is also critical. Install regular inspection and cleaning procedures for rubber molds. Keep workstations and wax trays clean and free of contaminants that could get in or on the waxes. While this type of contamination may not cause problems with the metal itself, it can cause surface defects, increasing the work time needed to obtain a level, smooth surface on the casting. In addition, because eliminating surface defects requires the removal of metal, it increases the rate of metal loss, and increases the possibility of working down to subsurface porosity, resulting in more repair work.

## Poor Seals on the Wax Tree

The wax tree can also contribute to defects. One of the most important issues here is the quality of the connection between the gate and the sprue. (The gate is the part connecting the jewelry wax pattern to the central trunk of the tree. The sprue is the central trunk.)

The gate/sprue connection (filet) should be sealed and smooth all the way around. Many people mistakenly use a wax pen to melt a hole into the sprue, then insert the gate into that molten spot. This method leaves sharp inward angles or points in the wax, which will turn into sharp brittle pieces of investment that can break off during casting. When they do, they are pushed to the outside of the mold cavity because of the difference in density between the metal and the investment.

To avoid this potential inclusion, the wax pen can be used to create a filet around the base of the gate/sprue junction. But to do this job faster and reduce the chances of a defect, dip the end of the wax gate into a small melting reservoir of sticky wax. (Thanks to Eddie Bell of The Bell Group in Albuquerque, New Mexico, who showed me this trick years ago.) You still need to seal around the base of the gate/sprue to create a smooth filet, but the sticky wax supplies some material to create the filet with. It also makes the tree building function much faster.

Be aware, though: The travel time between wax reservoir and sprue can allow the surface of the sticky wax to cool slightly. When this happens, unsealed gates sometimes snap off the sprue. To avoid this problem, pick up your wax pen once per row or column and seal the connections.

## Fixing Your System

Knowing that you have defects is easy; the challenge is in knowing where the defects really come from and how to fix them—especially when there are several stacked causes. It isn't enough to fix some of the problems. You must fix them all.

For consistent results, you must go through your system continually to reduce or eliminate all the possibilities that can cause these types of defects. The good news it that none of the prevention measures described in this article are difficult to implement. I have often observed that causes of the problems I am called in to solve are relatively obvious, and may have been noticed before. But because the defects have grown slowly over time, or the workers were too busy, the problem reached a critical point. The practices that are causing the problem have become "normal" and go unrecognized or forgotten. By questioning everything and forcing yourself to become aware of all potential problems, you can stop them well before they reach the rework stage.

# TRADE TIPS FOR Working with Wax

## Converting Wax to Metal

Many jewelers struggle with how to figure out just how much metal to use to cast a particular piece. It's actually a pretty straightforward process, says Gregg Todd, industrial training and project administrator at Stuller in Lafayette, Louisiana. He says there are three basic things that you need to know: the weight of the wax, the specific gravity of the metal to be cast, and the method of spruing for casting—i.e., button (in which patterns and sprues are built as a tree, with the main sprue forming a "button" at the end), pincushion (in which individually sprued patterns jut out from a convex mound), or doughnut (in which the main sprue takes the shape of a circular rim, onto which the patterns are attached). He provides the following advice:

"The first step is to multiply the wax weight by the specific gravity of the intended metal. The weight scale of choice doesn't matter as long as you stay consistent. If you use grams, everything will stay in grams. The same goes for pennyweights (dwt.). The table below lists the approximate gravity for various common metals:

| | |
|---|---|
| Silver: 10.53 | 18ky: 15.58 |
| Sterling: 10.4 | 18kw: 14.64 |
| 10ky: 11.57 | 18k palladium: 15.48 |
| 10kw: 11.07 | Gold: 19.36 |
| 14ky: 13.07 | Plat/Ir 90/10: 21.5 |
| 14kw: 12.61 | Plat/Ir 95/5: 21.5 |
| 14k palladium: 14.37 | Plat/Ru 95/5: 21.4 |

"Example: The wax weight is 0.67 dwt. and you are casting in 14k yellow gold: 0.67 dwt. x 13.07 = 8.76 dwt.

"Now include the spruing method.

Button: Multiply by 1.25 (8.76 x 1.25 = 10.95).

Pincushion: Add 20 dwt. (8.76 + 20 = 28.76).

Doughnut: Add 50 dwt. (8.76 + 50 = 58.76).

"Although this example would lead you to think that button spruing is the way to go for everything, let's look at another example, in which the wax weighs 15 dwt. and everything else remains the same: 15 dwt. x 13.07 = 196.1 dwt. If you use a button spruing method, the total weight would be 245.1 dwt. The pincushion method, on the other hand, would require 216.1 dwt. of metal, and doughnut spruing would require 246.1 dwt.

"Although not as accurate, you can also use the water displacement method for determining the amount of metal needed. This method uses volume instead of weight. Submerge the pattern in water in a graduated cylinder and note the water level. Then remove the pattern and add metal into the water until the same level is achieved. This will give you the same volume of metal as the pattern. This is a good technique to use when you have a hollow form (such as an insect) that will be cast as a solid."

## Getting the Lead In: Keeping Small Holes Clear When Casting

When casting bracelets and other items with small holes, it's often easy for the investment to clog the openings. Consequently, a jeweler has to either re-drill the holes in the final cast piece or re-cast entirely. To avoid this, Ricardo Basta Eichberg, president of E. Eichberg Inc. in Beverly Hills, California, suggests the following tip: Insert a piece of pencil lead (solid graphite, not a composite) in the holes before investing. Since the graphite won't burn away, it will protect the holes throughout the process, and the lead can be easily removed once the piece is cast.

## Get a Grip! Securing Pieces at the Bench

Kate Wolf of Kate Wolf Designs in Portland, Maine, offers the following advice for those creating wax models:

"If I am carving something tiny in wax, I will cut a long slice of the wax. I can then carve one end, while the rest of the wax serves as an extra-long handle. I'll also leave the wax model attached for as long as possible, finishing up the top, sides, and undercutting. Then I cut the model off and finish the back."

# Branching Out

## CHOOSING THE RIGHT PATH FOR METAL FLOW

BY GREGG TODD

Like so many jewelry manufacturing methods, spruing is rife with controversy. To some, this process of creating pathways for metal flow is no big deal: Simply create a hole in the investment and pour in molten metal. To others, it's a complex, make-or-break system to control the physical transformation of metal from a liquid to a solid.

Well, the purpose of a sprue network is simple: to allow for the flow of metal so product can be cast. How to best accomplish that, though, can involve a series of decisions about sprue diameters, angles, and patterns that can leave the mind reeling. [For a guide to terminology, see page 42.] Yet one thing everyone can agree on is that the sprue network is vital to successful lost-wax casting. Choose the right path, and it can eliminate many of the problems that lead to defective pieces.

## Growing Grains

To best understand the importance of spruing, you should first know about a process called "progressive solidification."

When precious metal changes from a liquid to a solid, it goes through a series of transformations. The first of these is called nucleation. During casting, as the metal gives up heat to the surrounding investment mold cavity, nuclei begin to form on the metal's outer surface.

As the metal continues to cool, the nuclei grow and create distinct grains. These grains form a dendritic structure, which continues to extend from the nucleus until it either contacts another dendrite or hits the metal's outer surface. Since the surface gives off heat, that's where the solidification begins, advancing to the center of the molten mass as heat continues to dissipate. The whole process is often referred to as "progressive solidification."

This phenomenon is extremely important in spruing for one simple reason: As metal changes from a liquid to a solid, it shrinks. As it shrinks, any metal that is still liquid is drawn from the center of the mass (i.e., the main sprue) to replace the diminishing volume. If all goes well, metal will be able to flow freely, and shrinkage porosity won't be present in the casting.

For this to happen, the sprue must be designed and positioned with progressive solidification in mind. In a properly designed sprue system, the sprue itself should be the last part of the casting to solidify; this will allow the network to continue to feed the casting as shrinkage occurs. Since metal will always freeze first in the thinnest areas, the generally accepted practice is to have the minimum sprue diameter at least 1 to 1.5 times greater than the thickest cross-sectional area to be cast.

Unfortunately, this doesn't always happen. One of the major concerns in the jewelry industry is the cost of materials. Materials that do not contribute to a piece in a remunerable way are viewed as an expense—and since sprues are not sold as part of a finished piece, they fall into this category. Because sprues of the preferred size are difficult to remove in finishing—during clipping they can lead to the distortion of a casting, and they produce large attachments that must be filed or sanded away—they conflict with the goal of keeping expenses as low as possible.

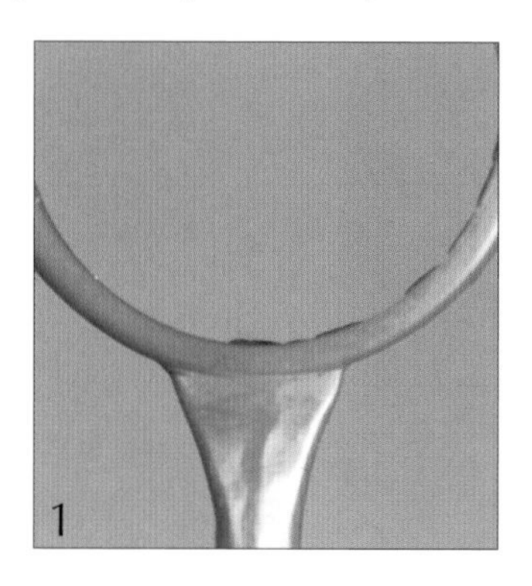
1

To make it easier for the finishers, some manufacturers design the sprue so it is both thinner and wider at the point where it attaches to the pattern (Figure 1). This opening offers the same cross-sectional surface area as the preferred sprue diameter. However, by thinning the sprue, you increase the rate of solidification, and the gate will often freeze before the pattern. This results in significant shrinkage in the area of the pattern below the sprue.

Since the purpose of the sprue network is to enable product to be cast, concerns about sprue attachment for finishing operations should be secondary to those for casting. So the question becomes: How can the spruing network best ensure a steady flow of metal?

## Branch to Branch

The first place to look is the design of the casting tree, the configuration most commonly used for high volume casting. A design may vary depending on the casting method; the feeder sprues of trees designed for vacuum casting often have a steeper angle than those used in centrifugal casting. The flask will also have some bearing on the tree's design; the most common bottom-pour flasks stand no more than 6 inches tall, while perforated chamber flasks can rise up to 10 inches. Obviously this will affect the tree's height.

Whichever method is used, the location and the configuration of the sprue attachment are the most critical elements contributing to casting success or failure. Unfortunately, most attachments are positioned and designed for processes other than casting, such as polishing (as discussed above) or the injection of wax or plastic patterns. Although the successful production of a pattern is important, the methods used for it do not readily transfer to the metal flow characteristics in casting.

The most common attachment location I see on rings is dead center at the bottom of the shank. This is about the best location for producing a rubber mold from the master, since it allows the mold maker to follow the sprue up to the pattern and create a smooth transition to the shank. It is also the location where the general linear flow characteristics of the metal are most evenly balanced. However, it's also the location where the ring shank usually has the narrowest and thinnest cross-section—the first place metal will freeze. This sets up a dilemma: Do we choose to maximize linear flow and set

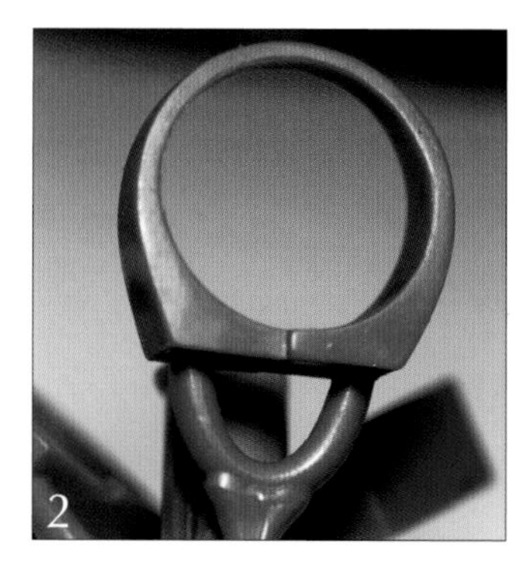
2

up the scenario for shrinkage, or to combat the consequences of solidification at the expense of flow? While the choice is ultimately the caster's, my experience has shown that it's easier to accommodate metal flow than to change the physics of progressive solidification.

The good news is that the location of a sprue attachment on a wax pattern does not have to be permanent. A sharp knife and a wax pen are all that's necessary for relocating it prior to building the tree. Generally, positioning the sprue attachment at the thickest cross-section of the pattern is most preferable (Figure 2). However, you must keep in mind the design of the piece. The head of a signet ring may have the thickest cross-section, but it may also contain a machine-cut pattern; by attaching a sprue, you've obliterated the pattern.

In such a situation, you may need to find alternate attachment sites, such as along the interior surface or below the design element. You may even need multiple sprue attachments to increase delivery to areas that otherwise wouldn't receive adequate metal flow. These alternative options may make the finisher's work harder, but they can reduce or eliminate remanufacturing costs by producing a sound casting every time.

Another consideration in sprue attachment is the metal you intend to cast. Here, one size does not fit all. Different metal types and karated compositions have unique properties, and the slush range between liquidus and solidus of a molten metal can vary. Generally, the higher purity metals have the smaller ranges, making them more difficult to cast (the greater the range, the better for casting). Sprue locations and designs used in the production of 14k yellow gold, for example, may not deliver the same results as when casting in 18k yellow.

Similar differences can be seen in different colored alloys, even if they're of the same karat: Yellow alloys, white alloys, red alloys, and green alloys do not behave the same. The reds, for instance, will generally not flow as well as the yellows, and the whites tend to be even more problematic in getting good fills (a consequence of the nickel in American metals). As a result, the sprue requirements differ for each, and a runner sprue may be required to ensure total filling. You'll need to experiment a little to determine the best approach for a particular metal/pattern combination.

## Great Gates

The gating design of the sprue attachment is another important consideration. Gates—those contact points where the feeder sprue connects with the pattern on one side and the main sprue on the other—need to provide a smooth passageway, so they should be free of all constriction points and sharp corners that cause abrupt directional shifts. (Figure 3 shows a sharp corner where the two feeder sprues connect.)

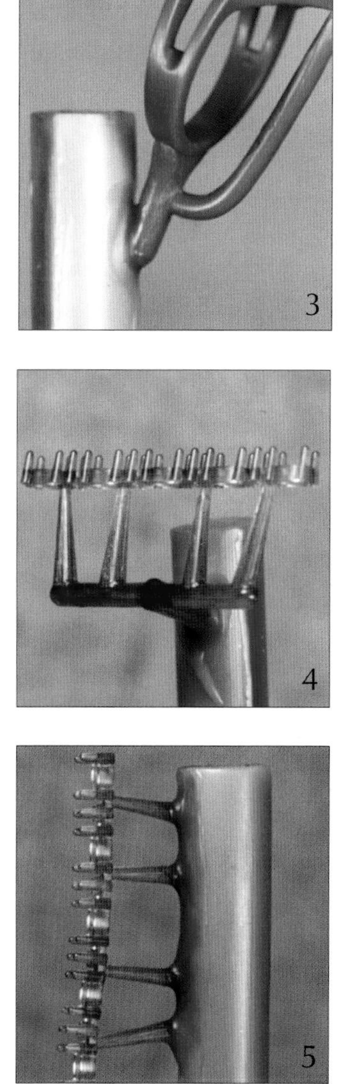


These factors will create turbulence, which in turn could lead to bits of investment being carried into the casting. Keep in mind that the sprue network is a system intended to transport a liquid to its final destination, the pattern. A gate should form a gentle curve, and it should never have a smaller cross-sectional profile than the feeder sprue itself.

You must also pay attention to the length of the feeder sprue. Remember, this sprue is not the Trans-Alaskan pipeline. It is not possible to pour molten metal down a narrow passage, surrounded by material that is 700°F to 1,000°F colder than the metal itself, along infinite lengths. For the oil to flow through the 700 miles of the pipeline, it must continuously be reheated along the way. This is not an option in casting.

In most cases, feeder sprues should range between 0.5 and 0.75 inch long. In those instances where longer feeder sprues are required, a larger sprue diameter should be used to help slow the solidification time and ensure proper delivery of the molten metal to the pattern.

Often, feeder sprues that branch to fill multiple areas of a casting are joined to the main sprue by a single conduit—a design configuration that results in excessively long feeder sprues (Figure 4). You should instead cut the feeder sprue where it branches and make multiple attachments to the main sprue (Figure 5). This both reduces the length of travel for the molten metal and conserves material. The tradeoff is that it may limit the number of pieces that can be placed on an individual tree, although that depends on the configuration.

You should also consider the angle of the feeder sprue from the main sprue to the pattern. As mentioned previously, this angle will occasionally differ, depending on the type of casting being done. For centrifugal casting, the angle should range between 45 and 60 degrees (the main sprue being a vertical 90 degrees). For vacuum casting, which relies more on gravity, the angle should fall between 25 and 45 degrees if you are using perforated flasks in a vacuum chamber. If you are vacuum casting on a tabletop unit, without a vacuum chamber encasing the flask, steeper angles should be used. Keep in mind that these angles are a general reference and will often vary considerably due to design considerations.

## Heading in the Right Direction

The directional flow of the metal cannot be ignored, either. Metal will flow in two primary directions: down and lateral. For centrifugal casting, think of down as the direction parallel to the main sprue (even though it may be mounted on a horizontal axis). When metal enters the flask, it runs along the main sprue until it meets an obstruction. Once an obstruction is encountered, the metal will then move laterally. This downward movement is the delivery direction for the metal, and the lateral movement is the fill direction. (Think of filling a glass of water: As soon as water running from a faucet hits the glass's bottom, the water spreads outward in all directions and the glass begins to fill. Similar flow behaviors happen inside the flask during casting.)

Gravity plays a significant role in the behavior of the directional flow, though how it can be applied differs between centrifugal and vacuum casting. With centrifugal casting, gravity's role is less evident, but it still must be considered. Let's say you have a tree made up of engagement sets; half of the rings have prongs, and the rest are smooth bands. Rather than randomly arranging the pieces, group the pronged rings together on one side of the tree, with the prong tips facing the main sprue rather than the flask walls (Figure 6). When investing, mark the side of the flask with the smooth rings. Later, when the burnout cycle has been completed, the flask can be loaded into the centrifugal casting machine with the marked side up: This will allow gravity to assist in pulling the metal downward during casting, ensuring the small prong tips will receive a better consistency of fill.

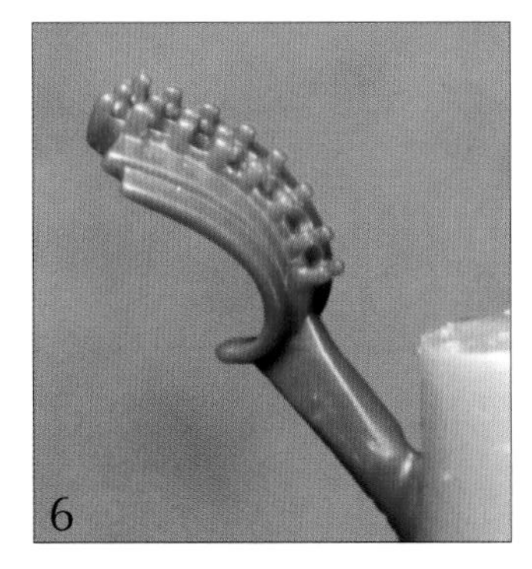
6

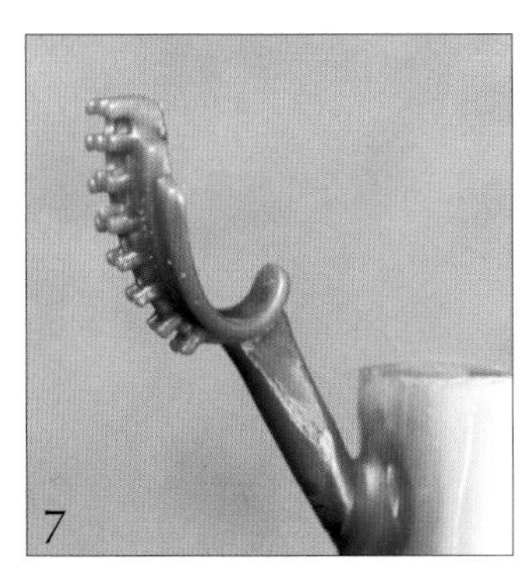
7

Since vacuum casting is done vertically, you must consider a different setup. The pieces should be arranged on the tree, first with gravity in mind, then with respect to the vacuum forces. Put the pieces with the finest detail and cross-sections toward the top of the tree, since this section will fill first and the metal will have the most fluidity. In addition, the vacuum forces will create a negative pressure inside the filled flask, helping to move the metal laterally. To best take advantage of the vacuum, position the prongs so they face outward, toward the sides of the flask (Figure 7).

## A Sprue by Any Other Name...

***A few definitions:***

The term "sprue" actually refers to either of two separate rods: the center rod ("main sprue") forming the trunk of the casting tree, and the connecting rod ("feeder sprue") between the pattern and the center rod.

"Gate" refers to two specific contact points: one between the center rod and the connecting rod, and the other between the connecting rod and the pattern. (A brief note: Some casters will refer to the entire connecting rod between the center rod and pattern as a gate. I choose to refer to it as a sprue, however, since it is applicable to other sprue configurations, such as a donut or pin cushion. [See page 37 for definitions of these configurations.])

The "sprue attachment" is the location of the gated connection between the connecting rod and the pattern.

A "runner" is a secondary feeder sprue attached to the pattern.

The "pattern" is the intended part to be produced.

The "base" or "rubber base" supports the entire tree assembly, which is referred to as the "sprue network."

## Building the Better Tree

Once you've addressed these considerations, building the tree is pretty straightforward. First, make sure you detail the patterns, making sure each wax is suitable for casting; apply solvent to smooth the surfaces and blend out defects. Next, separate the patterns by the type of metal in which they'll be cast, then by cross-sectional weight (not overall weight). Large cluster-top rings, for example, can often weigh more than a small solid signet ring, but the cross-sections of the parts constructing the cluster top are smaller than those of the signet ring. As a result, the cluster tops generally are placed at the top of a tree, which has a higher casting temperature than the bottom rows.

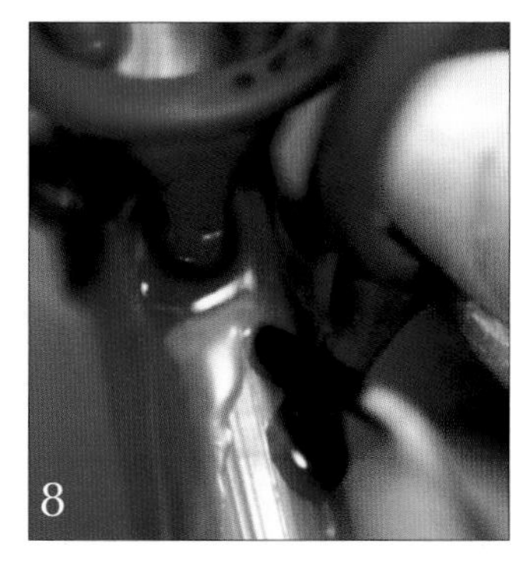
8

Pattern alignment is often a matter of choice. Some casters prefer to stagger the rows so they can pack the pieces more tightly. Others prefer a straight alignment, which causes less turbulence in the main sprue. Product design will often dictate which alignment style is most suitable. People clipping the patterns from the tree will also have opinions regarding which alignment is best, at least from their standpoint, but keep in mind that casting successfully is the primary concern.

As you build the tree, you'll want to minimize the risk of blowout. (Blowout occurs when the investment fails and the metal flows out through the opening.) Generally, you'll want at least 0.5 inch of investment covering the top of small trees with only a few patterns. On larger trees that will require more metal, consider having 0.75 inch at the top to reduce the chance of blowout. You'll also need to ensure you leave enough space between patterns; most casters consider 0.25 inch (6 mm) to be the minimum for the strength of the investment. In addition, there should be at least 0.25 inch (6 mm) between the pattern and the flask wall. Here the issue is less about strength than about thermal transference: After it's been removed from the oven prior to casting, the flask will begin to chill from the outer surface inward. If the outer surface chills too quickly, the pattern may not fill.

One of the biggest problems I've seen people encounter when building a tree relates to speed. Tree construction is a speed-and-rhythm process; more often than not, quicker is better. The feeder sprue to the pattern is usually cut to length with the wax pen, and a shallow depression melted in the main sprue. The feeder sprue is then positioned in the still-molten depression and tilted to the proper angle (Figure 8). However, to avoid gating problems, the small volume of wax pushed out around the end of the feeder sprue should be smoothed with the wax pen tip, creating a filet around the feeder sprue's base. (Excess wax can be removed from the pen tip by tapping it on a surface.) Since we have this need for speed, the wax pen will be set hotter than would be suitable for detailing a wax pattern. If the pen tip smokes slightly when wax is on it, it's probably close to the needed temperature.

Subsequent patterns can be sprued in the same manner, until you have a complete tree, ready for casting. As you can see, spruing will make a vital difference in producing defect-free pieces. You just have to keep in mind the sprue's basic purpose: to provide the quickest, least-restrictive flow of metal to a pattern. If you do that—and follow the guidelines above—you'll be on the right path to finding casting success. And that's something on which everyone can agree.

# Investment Advice

## HOW TO AVOID PROBLEMS WHEN INVESTING

BY PAUL COEN

More than 70 million pounds of investment powder is used worldwide each year by lost-wax casters in the jewelry industry. That adds up to well over $35 million spent for just one ingredient in the casting process. And when casters spend such serious money for powder, it is only fair for them to demand good, consistent results from their investment.

Yet all powders are not created equal. While the major investment manufacturers in North America, Europe, and Asia all aim for high quality, casters should understand that some powders are truly premium, while others are tailored for casting base metals. Ultimately, the investment's performance depends on the particular composition of ingredients, and on how well casters use—or, more important, misuse—their powder.

Gypsum-bonded investments, which are used for casting base

metals, gold, and silver, all contain varying amounts of gypsum, crystalline silica, and cristobalite, plus each manufacturer's special additives. Each contributes to the final casting:

- The expansion and contraction of the invested mold is determined by the percentage and quality of the investment's cristobalite content.
- Gypsum binder content is vital because the mold must be strong, yet offer easy breakout of the investment after casting.
- Crystalline silica in the powder should be of a specific particle size and shape for smooth surfaces on castings.
- Additives used to control setting time, air elimination, and how well the invested slurry adheres to wax patterns are vital to the quality of the final product.

So how do casters judge which powder has the best ingredients and composition for their applications? Should cost, supplier reputation, or past experience with the powder be the determining factors? Or is it just a shot in the dark?

Cost is not always a true indicator of quality, although the quality must be reflected in the price to some extent. Higher prices could be due to high quality ingredients; they could also result from the powder maker's high profit margin or the dealer's markup. (In most cases, investment powder is not a high profit item for either the manufacturer or the dealer; high prices usually stem from the rising cost of raw materials.)

Rather, the best way to judge the quality of various investment powders is by using and evaluating them in your operation. Stay objective, look closely at the casting each powder yields for you, and don't be afraid to sample other powders to see how well they perform against what you've used for years. Keep track of what it is costing you to finish your castings, and of your number of rejects. Are the dollars you're saving on a drum of less-than-premium powder greater than your potential losses, as polishers attempt to salvage less-than-quality castings?

That said, jewelry casters would do well to recognize that the plethora of defects suffered on an almost daily basis are normally not the result of the raw materials used. All jewelry casters seek the Utopia where all cast items have a smooth, fine-grained surface free of defects. That search leads them to look for new and easier methods; when casters visit trade shows, they ask investment suppliers, "What do you have that's new?" hoping that someone has come up with a panacea for their casting woes.

Instead of searching for a high-tech solution, they would be better off considering the faithful old basics, especially during the initial stages of casting—the mixing and handling of investment powder. Perhaps it's time for a review of some common investment-related defects that keep cropping up, what causes them, and some steps to eliminate them and make the handling of investment powders fail-safe.

## Avoiding Nodules (Bubbles)

Sure, the finishing department can get rid of nodules, but at what cost? Extra finishing time is wasted money, especially when some simple maintenance may solve your

### WATER-TO-POWDER RATIOS FOR INVESTMENT

| WEIGHT OF POWDER | 38/100 CC. | 38/100 OUNCES | 40/100 CC. | 40/100 OUNCES |
|---|---|---|---|---|
| 1 lb./0.45 kg | 173 | 6 | 182 | 6.4 |
| 5 lbs./2.25 kg | 862 | 30.4 | 908 | 32 |
| 10 lbs./4.5 kg | 1,725 | 60.8 | 1,816 | 64 |
| 15 lbs./6.75 kg | 2,588 | 91.2 | 2,724 | 96 |
| 20 lbs./9 kg | 3,450 | 121.6 | 3,632 | 128 |
| 25 lbs./11.25 kg | 4,312 | 152 | 4,540 | 160 |

problem at the source.

The first place to look when nodules show up is the vacuum pump on your vacuum table. The oil should be changed on a regular basis because it will get gunked up with water and investment slurry. If the oil needs changing, the pump cannot run at full capacity, and your vacuum gauge may be fooling you when it tells you that a full vacuum is being pulled.

(The best way to determine if the vacuum pump is working well requires only a few minutes. Place a half-filled glass of water on the vacuum table and cover with the bell jar. Turn on the pump: If the water hasn't come to a vigorous boil within approximately 30 seconds, you have identified a problem area.)

At the first sign of nodules, change the oil and establish a regular schedule for future oil changes. Better still, for shops casting 10 flasks a day or more, why not change the oil every Friday afternoon? It takes only a few minutes, the cost of oil is paltry, and that oil change takes a variable out of the game.

It also pays to periodically check the hose connections on the pump, and watch for wear on the rubber tabletop: this may prevent a good solid seal.

Once your equipment is in top condition, it's time to turn your attention to the investment. Always use the powder manufacturer's recommended water-to-powder ratio. Investment slurry that is too thick can inhibit a good vacuum.

Try running your investment mixer on slow speed during the entire cycle. The resulting slurry will be a good homogeneous mix, and you'll avoid pulling excess air into the investment that must be removed during vacuuming time. Also remember that when handled for too long, the slurry may begin to thicken and retard the vacuum process.

## Avoiding Water Marks on Castings

Water marks are best described as worm tracks, sort of a stringing effect. They result from investment slurry that is too thin or from procedures that don't use all of the work (handling) time recommended by the powder manufacturer.

Sticking with a ratio of 38 parts water to 100 parts powder is always safe for an investment mix. It will ensure a good surface finish and give a margin of error in case of slip-up on water measurement. (See the accompanying chart on suggested water/powder ratios.)

Every casting room should have a reliable scale for measuring powder, and guesswork on weight should never be permitted. When measuring the water for the mix, be sure to use a large beaker with clearly visible quantity marks.

When casters begin to judge the correctness of a slurry by feel, rather than by actual measurement, the headaches begin. Although many powders may have relatively wide parameters allowing for mistakes, why gamble? It is not unusual to see a worker reach into the slurry and pinch some of it between his thumb and his forefinger, then decide it needs another couple of ounces of water. That may have worked for grandmother when she whipped up a cake, but is that really the best approach for turning precious metal into salable jewelry?

Work time begins when the powder is added to the water, continues through mix and vacuum time, and ends when the flask is placed on the workbench to set. Some powder makers tailor their product to a work time of 8 to 9 minutes, and others have a work time of 9 to 10 minutes. There is usually a variance of about 30 seconds.

If your mixer doesn't have a timer, buy a separate timer and insist on its use. Guesswork on mix or vacuum time is common, but can be fatal. Remember that water-rich investment is drawn to the surface of the wax and can cause erosion and particle separation.

Most powder makers suggest a water temperature of 70°F to 75°F (21°C to 24°C). Colder water will increase the available work time of the slurry, and warmer water will shorten the work time, but for best results you should follow the water temperature suggested by the manufacturer.

When not using the powder, keep the inner bag closed and the drum sealed.

## Avoiding Finning and Flashing

If there are casters who have never encountered this problem, it is because they are rare individuals who never, ever take shortcuts. This defect on castings gives finishers ulcers, but it can be avoided by sticking to established procedures.

Finning and flashing occur when too much water is added to the investment mix, severely weakening the mold. In this case, the mold simply cannot stand up under the shock of molten metal entering the cavity. Again, stay within the powder maker's suggestion on the water-to-powder ratio, and do not continue to handle the invested slurry beyond the manufacturer's recommended work time.

Once the final vacuum is pulled, the invested flasks should be set on a bench and remain untouched. Small flasks should set for at least one hour, and larger ones of 5 to 6 inches for at least two hours. Keep in mind that the top of a flask will harden early on, while inside, the slurry will not have completely firmed up.

Mishandling the flask by pulling the rubber base, removing tape, or cleaning off excess investment before proper bench set time can cause the sprue and patterns to move inside the walls of the mold, resulting in finning and flashing, plus poor surface finish.

Avoid the practice of placing invested flasks on rolling carts and then moving them about. Give the mold ample time to firm up, but be careful not to leave them for an excessive period of time, either. Some casters invest their flasks on a Friday afternoon and then allow them to bench set until placing them in the furnace late on Sunday. The flask will lose all of its moisture during that period, adversely affecting results. Cover the flasks with wet towels when they set for that long.

## Avoiding Mold Blowout

To help avoid this problem, pay close attention to the water-to-powder ratio in the investment mix. Too much water weakens the strength of the mold, and the molds cannot stand up to the force of the molten metal. Also, do not pack too long a sprue in the flask; although this may seem cost efficient, an inadequate amount of space between the investment and the tree can permit the molten metal to blow right through. [See also "Successful Investing," page 50.]

## Establishing Proper Procedures

The best approach to investing is to always attempt to do it right the first time. Establish a non-negotiable standard operating procedure for all phases of the investing operation, and check on a regular basis to be sure those procedures are being followed.

Insist on a clean work environment. Casters who do platinum as well as gold casting must be careful to keep the powder used for platinum alloys away from gypsum investment. Platinum powders are usually phosphate bonded and can contaminate gypsum-bonded investment powders, affecting their work time. Clean all equipment, including mixing bowls, after using platinum investments. Do not store the different powders together, to avoid the chance of human error.

A clean work environment will also contribute to the safety of the workers. Investment powder, handled properly, is safe, but understand that cristobalite is a suspected carcinogen in humans and that continued inhalation of crystalline silica can cause silicosis.

Casting manufacturers should have in place an adequate ventilation/exhaust system that eliminates powder inhalation by workers. The microscopic particles cannot be seen by the human eye, but they are present. Workers can also protect themselves by wearing respirators while working around the powder.

In addition, you should read the warning label on the investment drum, obtain Material Safety Data Sheets from suppliers (not just on powder but on all the materials you purchase), seek assistance from the appropriate government agencies, and consult with firms that are able to assist you in cutting down on employee risk. It may pay to look at equipment that enables casters to mix and pour investment under a vacuum. This equipment reduces dust levels dramatically and eliminates the possibility of human error by allowing for accurate mixing, vacuuming, and pouring.

Bad work habits often creep into the early stages of casting and lay the groundwork for problems. When everything runs smoothly for weeks, then rejects suddenly begin to pop up, it's a safe bet that shortcuts were employed. The key to getting the most out of your casting investment is to get rid of the shortcuts and poor casting practices, and to work diligently on the fundamentals. By selecting the proper powder and following basic procedures, casters will find their investment producing good, consistent results—and that's the difference between increased profits and a dismal bottom line.

# TRADE TIPS FOR Investing Flasks

COMPILED BY SUZANNE WADE

## Weighing In

You know it's going to be a really bad day when you break out your flasks and the castings look terrible—especially if you're certain you haven't made any changes to your process.

If your methods haven't changed, it's possible your equipment has. Many pieces of equipment require regular adjustment to maintain their accuracy, so one of the first steps in the troubleshooting process is to check the calibration of your equipment.

If your investment is cracking, flaking, or breaking down—problems that can be caused by an incorrect water-to-powder ratio—check your scale, suggests Daniel Grandi of Racecar Jewelry Inc. in Cranston, Rhode Island. "Postal scales and investment scales can easily lose their accuracy by as much as a pound," he observes, causing the caster to use too much or too little investment powder.

One way to calibrate the investment scale is to place scrap metal, such as nails or chunks of brass, into a plastic Ziploc bag and weigh it on a small digital scale. Add or subtract metal pieces until you reach exactly the same weight as the amount of investment powder you typically use, then zip the bag shut.

Place the plastic bag containing the metal on the investment scale. If the scale doesn't show the correct weight, you can adjust it using the small knob or screw most scales have for this purpose.

Another possible source of problems is the burnout oven. Since burnout requires a series of steps at specific temperatures, if the oven's temperature controller starts misbehaving, it can spell big trouble for the caster. Incorrect calibration can be the cause of two problems: If the temperature is too low, waxes won't burn out completely, and if the temperature is too high, the investment can crack and spalling can result.

To check when and whether your oven is reaching the correct burnout temperatures, get some calibrating cones from a pottery supply house, suggests John Henkel of J.A. Henkel Co. in New Brunswick, Maine. "These melt or 'slump' when they reach the temperature written on their sides," he explains. By checking the temperature shown on the gauge at the moment the cone slumps, casters can determine whether their oven's pyrometer is accurate.

## Stop the Shrinkage

Here's a small-shop trick from Craig Bowlby at Bowlby Arts in Atlanta for minimizing wax shrinkage during burnout.

Immediately after pouring and vacuuming your flask, submerge it completely in a container of warm water. Don't worry about the water diluting your investment: Submerge the flask slowly, and that won't happen.

The heat from the water will penetrate the investment and heat your wax model, causing it to expand. It will also accelerate the cure of the investment. Once the investment has cured completely, do a normal burnout. This should give you a casting with very little shrinkage, if any.

## Fast Delivery

Even experienced casters occasionally have problems with investment cracking on flasks. The cracking is usually caused by not moving the invested flasks to the drying table fast enough.

Since you have only several seconds to move flasks from the investing table to the drying table, it can be a challenge to make the fast delivery if you invest several flasks at a time. Daniel Grandi of Racecar Jewelry Inc. in Cranston, Rhode Island, shared his innovative solution to this problem with the Orchid e-mail forum.

"We invest six 4 inch by 7 inch flasks at one time," he wrote. "We had problems moving the flasks off the machine quickly enough—we have about 15 seconds to move all six." The solution, Grandi figured, was to move all six at once. To do so, he bought several heavy plastic pizza pans that serve as trays.

"I cut away an area of the pan where the pipe goes through the top of the investing table," he wrote. "We put all six flasks on the pizza pan, and when the mix is ready to pour, we move the pizza pan with all six flasks onto the table."

When investing is complete, the flasks are whisked off the investing table and set on the drying table in about five seconds—a time that would make any pizza maker proud.

# Successful Investing

## FIVE STEPS YOU SHOULDN'T FORGET

BY GREGG TODD

### Measure the Height of the Tree before Investing

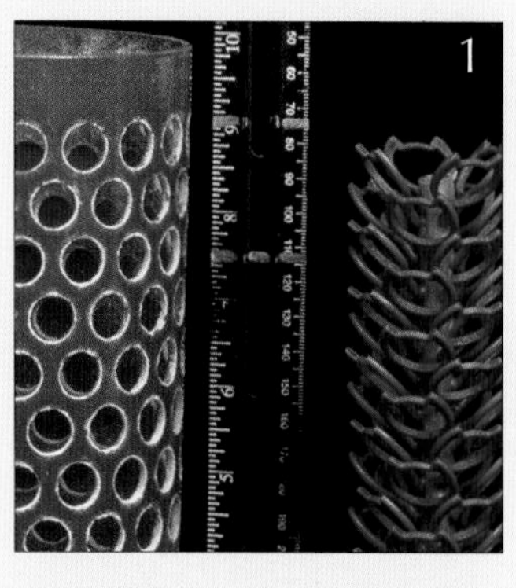


A good practice for preventing investment blowout is to measure the height of the trees before investing (Figure 1). The maximum tree height should be between 0.5 inch and (for larger trees with more metal) 0.75 inch shorter than the fill level of the flask. It's also important to leave a 0.125 inch gap between the top of the investment and the top of the flask if you are using benchtop vacuum casting with a solid flask. With perforated flasks cast in a vacuum chamber or with centrifugal casting, the flasks can be filled completely and the tree height lengthened accordingly.

### Dip the Tree in a Debubblizer

Unless you are mixing the investment and filling the flasks in a vacuum, dipping the tree in a debubblizer is recommended (Figure 2). This reduces the surface tension on the patterns, which helps to prevent air bubbles from clinging to the surface and forming nodules on the casting.

### Properly Position the Tree in the Flask

How the tree is positioned in the flask for casting is very important in ensuring investment strength. There should be ample space—generally a minimum of 0.25 inch (6 mm)—between the outermost part of the pattern and the side wall of the flask (Figure 3).

The positioning will also help you to calculate the amount of time you have before the flask has cooled and your castings are in danger. Most gypsum-based investments have a heat transference rate of about 400°F (204°C) per minute/mm. If the investment surrounding the tree is 6 mm thick, you will have about 1.5 minutes before the temperature at the outermost portions of the pattern drops 100°F (38°C). If you have a 200°F (93°C) window for casting, you will have about 3 minutes to complete the casting before the temperature drops below that limit. If you have 12 mm of investment between the pattern and side walls, you will have about 6 minutes to complete the casting.

### Carefully Measure and Monitor the Investment

When it comes to investing, following manufacturers' recommendations for mixing and measuring is critical. Powder should be measured by weight and fluid should be measured by volume.

Safety is a consideration when mixing investment. You should wear a good particle mask that seals tightly to your face. Wearing a paper dust mask is only marginally better than wearing no protection at all; it provides a false sense of security.

4

After carefully measuring and achieving the correct water-to-powder ratio, the investment must be mixed thoroughly to form a creamy slurry (Figure 4). Carefully follow the manufacturer's directions for mixing times and temperatures. Keep in mind that the water and powder temperatures will affect mixing time. The hotter the materials, the shorter the mixing time.

Note: Investment has a shelf life. Over time—between six months and one year—the investment can absorb humidity from its surroundings and become ruined. It's important to store investment according to the manufacturer's instructions. One tip: Never store investment directly on a concrete floor, since concrete holds moisture.

Aside from problems with humidity, the silica begins to settle out of the investment powder after a while. If not re-mixed periodically, the investment formula can change enough to cause major problems in casting. The most common problems associated with this are fragile investment and poor surface texture on castings.

## Vacuum the Investment

The mixed investment must be vacuumed to remove trapped and suspended air. When vacuuming is complete, carefully pour the investment into the flask and fill it to the top of the tree. Place a rubber sleeve over the top of the flask and vacuum the flask (Figure 5). The investment will rise and boil during vacuuming as the trapped air expands and rises. After vacuuming, add a little more investment to top off the flask to the proper level, as described earlier, and set it aside to gloss off (solidify).

Before loading the flasks into the oven, care should also be taken when removing the rubber. Improper handling can cause small amounts of investment to chip off the surface, thereby causing inclusions in the casting.

5

# Toward a Better Burnout

## DISCOVERING WHAT HAPPENS INSIDE THE KILN

BY JOHN MCCLOSKEY

As manufacturers well know, a piece of jewelry is only as good as the materials and methods used to produce it. That's especially true with lost-wax casting; everything from wax selection to investment mold temperature can determine whether a cast piece is smooth and shiny—or riddled with porosity.

Fortunately, manufacturers can find a range of information, from the flow characteristics of alloys to the properties of various investments, to help them select the best materials and use the proper methods. Unfortunately, there is still one area of the lost-wax process for which little information exists to answer a vital question:

What happens to a gypsum-bonded investment casting mold when it goes through the burnout cycle?

For casters, such knowledge can be crucial: It can prevent the mold from cracking and the casting from being ruined. But it can also be difficult to obtain. After all, of all the steps necessary to prepare a mold for filling, the burnout of wax is the one that remains invisible. Carving a wax pattern, creating a rubber (or metal) mold, mixing investment, spruing a tree, melting the alloy, preparing a flask—for all of these steps, casters can immediately inspect their work. The burnout of wax, however, happens in the hidden cavities of the set investment; if problems occur during this phase, they will become apparent only after the tree has been broken out.

The key, then, is to know what to expect *before* the flask is placed in the oven. The caster must gain insights into how high temperatures will affect the investment, and how those temperatures should be best applied. They must also know the chemical reactions occurring as the wax or plastic burns away, and their effects on the final cast piece.

Toward this end, I've given below some insights into the burnout process, as well as observations from my own study of a typical burnout cycle in a gas-fired oven. (Note: The results apply equally to electric ovens, the only difference being that electric ovens don't have venting around a hearth plate, so more attention needs to be paid to door venting and vents on the top of the chamber.) Throughout this study, I evaluated the changes to the investment molds, noting their conditions at various points in the process. I also compared the different results between wax and plastic patterns.

Ultimately, these detailed observations helped explain why burnout processes take so long, and the reasons behind a mold's cracking—as well as how to prevent such a problem.

## An Oven Overview

The first step toward better burnouts is to have a working knowledge of the actual burnout oven—what it is, what it does, and how it works.

Gas-fired ovens basically accomplish three things: They dry and cure the investment mold; they melt, burn, and remove pattern materials (i.e., wax or various plastics) from the mold's cavities; and they heat the mold to a proper casting temperature. Typically, a well-built oven will consist of little more than a few simple elements:

1. A steel box.
2. A 100,000 to 125,000 Btu/hr. gas burner mounted underneath the hearth plate.
3. A thermocouple to regulate the heat.
4. A microprocessor-based controller to program specific temperatures and their durations.
5. A thick layer of lightweight insulation to minimize heat loss. (A lightweight lining also has the advantage of cooling quickly at the end of a production run, allowing another load of wet molds to be quickly loaded for the next casting cycle.)

Just as the basic parts of a burnout oven are relatively simple, so too is the way the oven works. Typically, the gas burners operate through the Venturi mixing of natural gas and combustion air. In this method, the flow of the gas naturally creates a vacuum that draws the combustion air into the body of the burner. The gas and air combine and travel toward the many small burner tips located along the burner and beneath the hearth plate in the oven. The mixture is then ignited by either pilot light or ignition spark.

However, when the oven ignites, more than just hot air travels into the oven chamber and comes into contact with the molds. This is when the process gets trickier, and it becomes important to know something about the chemistry of the burner flame, the need for both primary and secondary supplies of combustion air, and how small variations can influence the burnout process.

First, natural gas is mostly methane ($CH_4$). Under conditions of complete combustion, it reacts with the two parts of oxygen ($2O_2$) and the eight parts of nitrogen ($8N_2$) in the combustion air, producing the following formula:

$$CH_4 + 2O_2 + 8N_2 \longrightarrow CO_2 + 2H_2O + 8N_2$$

In essence, it's not just air now in the chamber, but also water vapor ($H_2O$) and carbon dioxide ($CO_2$). And if the natural gas in the burner doesn't combine with sufficient air, then carbon monoxide may be present. In fact, the only element left unchanged in this reaction is the nitrogen, which is not "burned" but is instead heated to the temperature of the natural gas flame, which in turn helps to transfer heat to the mold.

Now, in addition to the primary combustion air needed to supply the gas burner, an oven needs a secondary supply of air for burning the wax or plastic patterns in a mold. This secondary air comes into the oven through the burner slot, underneath the hearth plate, and around the oven door. The oxygen in this secondary air is the main oxidizing agent for burning the wax or plastic pattern residues, which are carbon-based.

As these residues burn away—adding to the water vapor, carbon dioxide, and possible carbon monoxide already in the chamber—the oxygen in the secondary air will be diluted significantly: Where normally the air contains 20 percent oxygen, now it contains only 10 to 12 percent. The less oxygen contained in the oven chamber, the greater the chance of producing molds with carbon residues in the investment.

There is no simple control valve to regulate the amount of secondary air that enters a burnout oven. However, there is a simple way to introduce additional oxygen into the air: Increase the amount of air in the *primary* supply.

Let me explain. The primary air requirement for complete combustion is not significantly large: When one cubic foot of methane is burned, 1,000 Btu of heat is released, and 10 cubic feet of primary combustion air must be supplied to the burner. Therefore, if a burner is rated at 100,000 Btu/hr., it can use 100 cubic feet of natural gas and 200 cubic feet of air in one hour of continuous operation.

But let's assume a burner is adjusted so that it receives twice as much air as required for complete combustion. Under these conditions, the reaction becomes the following:

$$CH_4 + 4O_2 + 16N_2 \longrightarrow CO_2 + 2H_2O + 2O_2 + 16N_2$$

You still get carbon dioxide and water vapor, but you now have additional oxygen and nitrogen in the chamber. The same 1,000 Btu of heat will be generated when one cubic foot of methane is burned; the additional gases will absorb some of that heat and have a slight cooling effect on the flame.

However, this oxygen-rich flame can benefit the caster: The chamber now contains hot oxygen at the flame temperature of the burner, which will create an atmosphere where the patterns—along with any carbon residue generated by their decomposition—can be more efficiently burned. (Carbon does not evaporate during the burnout process; rather, it must be burned with oxygen from air flowing through the oven.) By maximizing the amount of oxygen inside the oven, a caster can better ensure that clean, carbon-free molds will be produced.

Understanding the effects of the combustion process is only the first step. To achieve maximum success, the caster must also understand the characteristics of wet investment molds and how the molds are dried during the burnout process.

## Wet and Dry

Investment molds consist of a powder composed primarily of gypsum, silica, and cristobalite. This powder is, according to the powder manufacturer's recommendations, blended with water to form a creamy slurry. Again, it sounds simple, and it usually is. Casters just have to make sure that all the water is removed from the mold during the burnout process, without damaging the mold itself.

The calcium sulfate binder used in casting investments causes two forms of water to be present in molds: "chemical" water and "free" water. The chemical water (or water of hydration) becomes chemically bound to the gypsum ($CaSO_4 \bullet 0.5H_2O$) while molds harden:

$$CaSO_4 \bullet 0.5H_2O + 1.5H_2O \longrightarrow CaSO_4 \bullet 2H_2O$$

This chemical water makes up about 13 percent of the water found in the mold. (A typical 4-inch-by-8-inch investment casting mold will initially contain more than 1 lb. of water overall.) The balance of the water used to mix the investment—the free water—resides in the many pores common to investment. Investment is highly porous when it hardens, providing plenty of space for water to collect. And the only way to remove this free water is to boil it away, as if it were in a pan on a stove. The porosity allows the resulting steam to escape easily. As long as free water remains in the porous investment, the temperature of the mold will be arrested at the liquid's boiling point of 212°F (100°C). (Water cannot be heated above its boiling point unless significant external pressures are applied.)

Removing all the water is another story. Breaking the chemical bond between gypsum and its water of hydration requires energy, which in turn requires higher temperatures. In fact, gypsum dehydrates in two steps. At 262°F (128°C), the following reaction occurs:

$$CaSO_4 \bullet 2H_2O \longrightarrow CaSO_4 \bullet 0.5H_2O + 1.5H_2O$$

The water that was chemically bound to the gypsum detaches. Then, at the higher temperature of 325°F (163°C), all the water found in the investment is removed:

$$CaSO_4 \bullet 0.5H_2O \longrightarrow CaSO_4 + 0.5H_2O$$

So molds will not be totally dry until they are completely heated above 325°F (163°C). Sounds simple. But how much heat can be applied at one time? Could a rapid increase in temperature also speed up the burnout process, or would it endanger the mold? To determine these answers, I decided to conduct my own tests.

## Investigating Burnout

The tests began with the assembling and investing of two trees: one with wax patterns, the other with Epolene plastic patterns set up on a wax sprue post. After the investment hardened, each mold was weighed and placed in a gas-fired burnout oven, with care taken to ensure there was plenty of oxygen for the secondary supply of air. A typical time-temperature sequence was used to control the oven temperature:

Ramp to 275°F (135°C) in 30 minutes.
Hold at 275°F (135°C) for two hours.
Ramp 275°F to 1,350°F (135°C to 732°C) in 4.5 hours.
Hold at 1,350°F (732°C) for four hours.

During the first seven hours, the molds were removed at one-hour intervals and weighed, and their temperatures were measured by a thermocouple inserted into the bottom of the sprue post. At each weighing, the oven temperature was also recorded.

Figure 1 contains plots of the mold temperatures and oven temperatures recorded at each interval. These plots demonstrate how the mold temperatures lagged behind oven temperatures at all times during the seven hours of data collection. During the early stages of the burnout, the flask temperatures were arrested at 212°F (100°C) while the free water remained in the investment; four hours passed before the molds climbed above this temperature, even though oven temperatures were increasing to about 600°F (316°C) during the same period.

Figure 2 is a plot of the weight losses observed in molds at each one-hour interval. During the first two hours of burnout, flask weight loss increased from zero to about 120 dwt./hr., while mold temperatures increased to about 180°F to 200°F (82°C to 93°C). Weight losses subsequently decreased to about 50 dwt./hr. and increased to a maximum observed value of about 200 dwt./hr. at the end of four hours. After four hours, mold weight losses continually declined.

The two peaks in weight loss appear to correspond to maximum losses in wax and free water. The first peak at the lower temperature indicates that wax/plastic is rapidly draining out of the molds during the initial stages of burnout. As mold temperatures remain at 212°F (100°C) for extended periods of time, the flasks' weight loss increases as the free water converts to steam. Mold temperatures increase once the bulk of the wax/plastic and free water is removed, and weight losses decrease as the molds become cleaner and drier. This is particularly true after about five hours, when the flask temperatures start to exceed 500°F (260°C).

At the end of seven hours, the molds were cooled and broken open to reveal their internal conditions. The mold that had con-

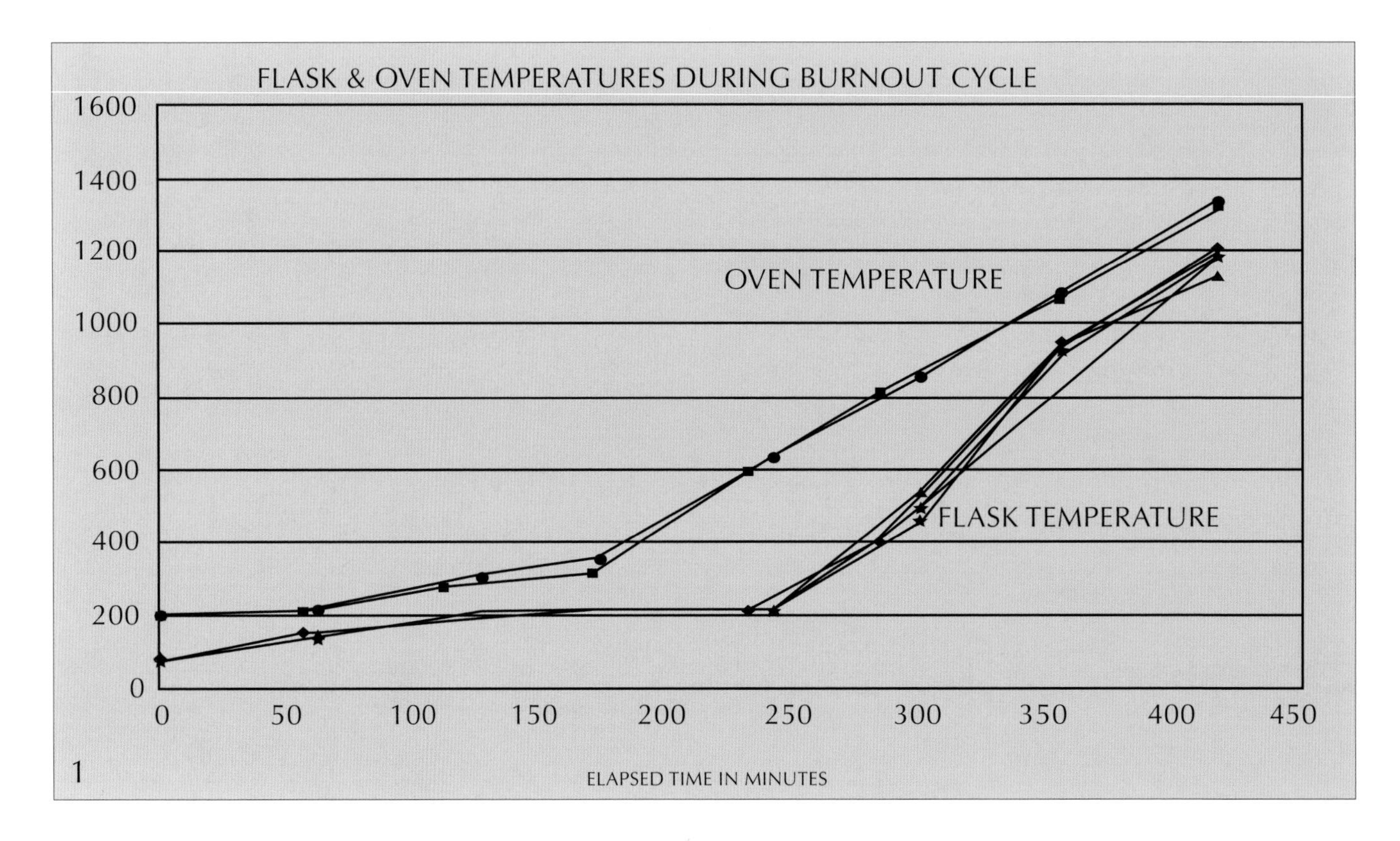

tained wax patterns had empty cavities surrounded by gray halos of wax residue that had been absorbed by the investment; obviously, seven hours of burnout processing had not been enough. To achieve full burnout, the molds would have had to remain in the oven for more time at higher temperatures.

The other mold (with Epolene patterns) contained a dark halo of carbon adjacent to the sprue post. The investment surrounding the cavities appeared to be cleaner and free of carbon. Apparently little or no residue from the Epolene had been absorbed by the investment. I attributed this to the higher viscosity of the plastic compared to the wax.

These observations helped explain to me why burnout processes take so long, as well as how molds can crack. The weight loss measurements indicate that it takes up to four hours for steam to escape during the low temperature stages of burnout. Even though the molds are highly porous, there is a limit to the rate at which steam can escape from the interior of a mold. It's a fact of casting that must be respected.

This led to the question, What if oven temperatures are increased too rapidly during the early stages of burnout? Thinking about it, I realized that as the temperature difference between the mold (which is fixed at 212°F/100°C) and the oven increases, heat is transferred into the mold at a faster rate. As this heat transfer quickens, so too is steam generated more rapidly in the mold.

However, if this increased volume of steam cannot escape from the mold—and, judging from the experiments, it cannot—internal pressure in the investment can build up to a point where the investment fractures. This steam pressure will be trying to push the investment out the top and the bottom of the steel flask, so fracturing will occur parallel to the top and bottom and appear to be horizontal across the mold cross-section. These horizontal fractures (also known as umbrella fractures) ultimately lead to ruined castings.

This problem can also be aggravated by the amount of cristobalite used in investment blends. Cristobalite expands significantly at around 518°F (270°C). If a rapid rise in temperature creates excess steam pressure while the mold is going through its cristobalite expansion stage, cracking can occur more readily.

These, then, are the keys to a better burnout: plenty of oxygen for the secondary air supply, investment without a trace of water, and adequate times at elevated temperatures—with an emphasis on duration. After all, there are limits to the rates of water and carbon removal that can be achieved safely, no matter how permeable an investment mold may be. But if those limits are respected, then burnouts can produce clean molds—and clean castings.

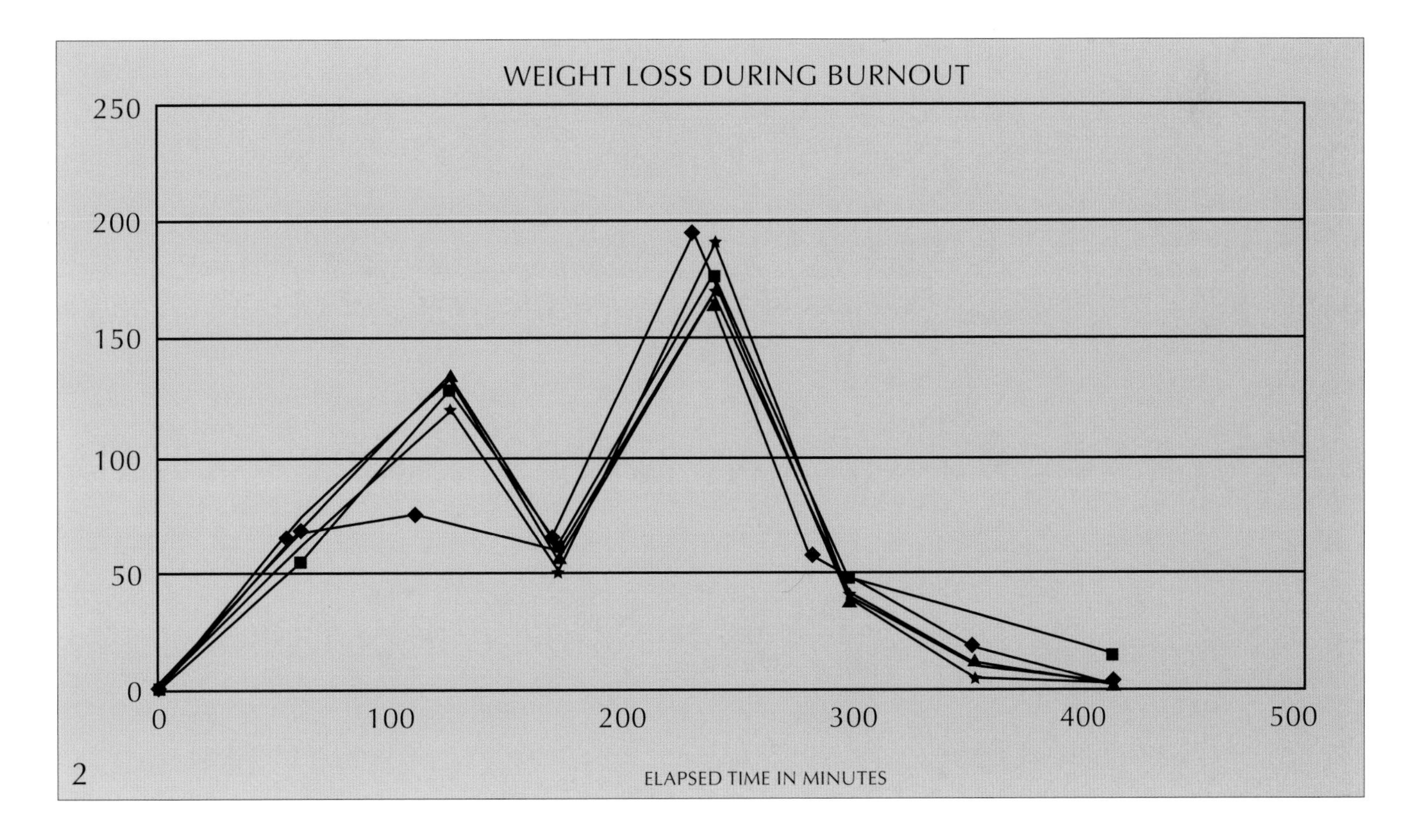

# Of Melts and Methods

## MELTING TECHNIQUES AND CASTING EQUIPMENT

BY DANIEL BALLARD

After burnout, you're ready to cast. And to cast successfully, you need to look at a few key factors, which I call MMCC: Metal used, Melt method, Casting method, and Capacity needed. These are the "bare bones" of the process, so to speak.

First, what metals are you casting? Gold, sterling, and other metals used for models and samples (in fact, all metals other than platinum) enjoy the ability to be cast in the same equipment, since the maximum melting temperature will be 2,100°F (1,149°C) or lower. However, if you're casting platinum—for which the melting temperatures can reach 3,200°F (1,800°C)—you will need special high-temperature equipment.

Next, how will you heat the metals up to those casting temperatures? You basically have two choices: flammable fuel gases with a torch, or electricity. If you're using a torch, be careful which fuel gases you choose, especially if you're working with platinum.

Acetylene is a dirty gas—it's what they use at the local muffler shop to weld steel—and a poor choice for jewelry work (especially platinum). Propane, hydrogen, and compressed natural gas are all suitable choices. Natural gas is the most commonly used gas in small- to medium-capacity shops. Hydrogen is the cleanest of the gases, and very hot—a hydrogen/oxygen combustion can approach or exceed 6,000°F (3,316°C)! Propane burns fairly cleanly, but it needs plenty of oxygen to boost the heat (it's also extremely heavy and explosive).

Beware of welding-shop branded gases, such as "Safe" or "Mapp" gas. These are welding supplier mixes and can vary. Ask what is in any gas with which you are not familiar. All gases have their peculiarities in regard to safety issues and the details of oxygen/fuel gas balance. Also be aware of your local ordinances; some gases must be permissible according to safety codes.

If you want to heat with electricity, you have two choices for gold and sterling: Resistance or induction. All resistance melters use some kind of high-temperature wire (like the wire in light bulbs) that is coiled and positioned outside the crucible. Electricity is fed through the wire, the wire offers resistance to the electricity, and the wire becomes hot enough so that heat radiates to the crucible and casting metal. The maximum temperature you can depend on with resistance is about 2,100°F (1,149°C). For platinum, resistance *is* futile.

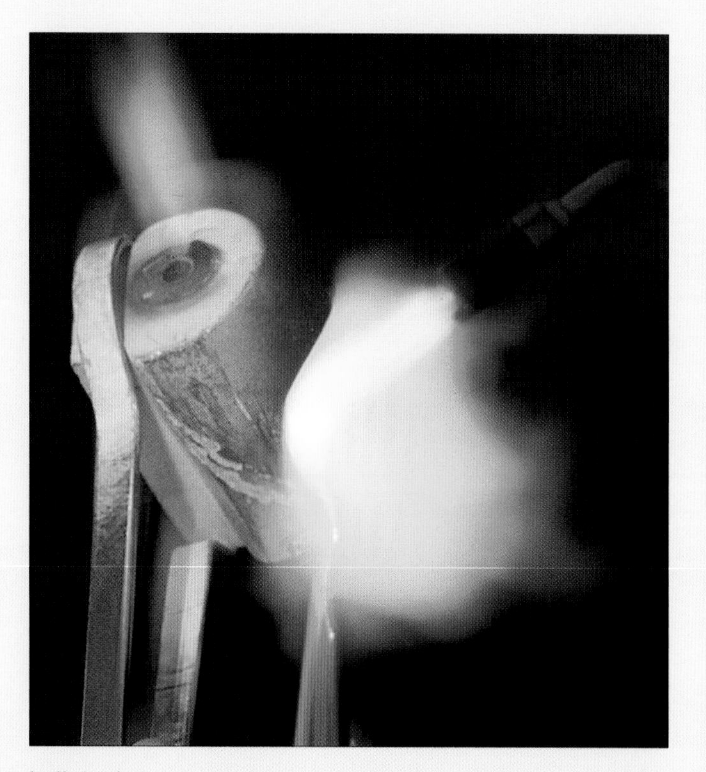

Induction heating is analogous to the microwave oven at home. It uses radio frequency energy at medium frequencies (6 to 12 kilocycles) or high frequencies (300 to 500 kilocycles) to "induce" heat directly. Different frequencies produce different results—for example, medium frequencies offer the major advantage of actually stirring the molten metal. Be careful when you buy this technology for platinum, since only induction equipment specifically designed to reach the high melting temperatures of platinum will do. Induction also needs lots of electricity, and a cool water source to keep it from melting its own parts down to a puddle.

The next question is how you should actually cast or throw the metal. Here, you have two ways to go—the famous spin casting (a.k.a. the centrifugal or angular momentum method) or vacuum casting. Spin casting, as the name says, spins the flask at a suitable rate to force the molten alloy into the flask. This is a very powerful method that allows very small parts to fill, and is capable of doing platinum as well as gold and sterling. Most common are the horizontal spin casters, though there are also vertical spin casters designed for platinum.

Vacuum is a method through which ordinary gravity is used to vacuum out air from under the perforated flask and to help move the air and gases from under the incoming casting alloy. This method is now very common and well understood. In a way, it's really atmospheric pressure casting. The latest high-capacity computerized casting machines actually provide pressure over the flask and vacuum under the flask. ("Pressure over vacuum" or "POV" is a major breakthrough for large-capacity stone-in-place production.) Vacuum casting has become very common these days—even very small shops often go with it. However, if you're casting platinum, vacuum alone will not ensure complete fills; you will need a machine that also adds centrifugal force.

The last factor you'll need to consider is raw capacity. Are you going to cast a few flasks per week or 200 per day? Will they measure 4 inches by 4 inches or 8 inches by 6 inches? There are many types of equipment available, all with varying degrees of capacity and speed; research your options appropriately to ensure you don't overspend for more capacity than you need, but also don't end up with a machine that can't meet your volume demands.

So remember: MMCC. These keys will open the doors to the correct equipment and practices, and help you on your way to successful casting.

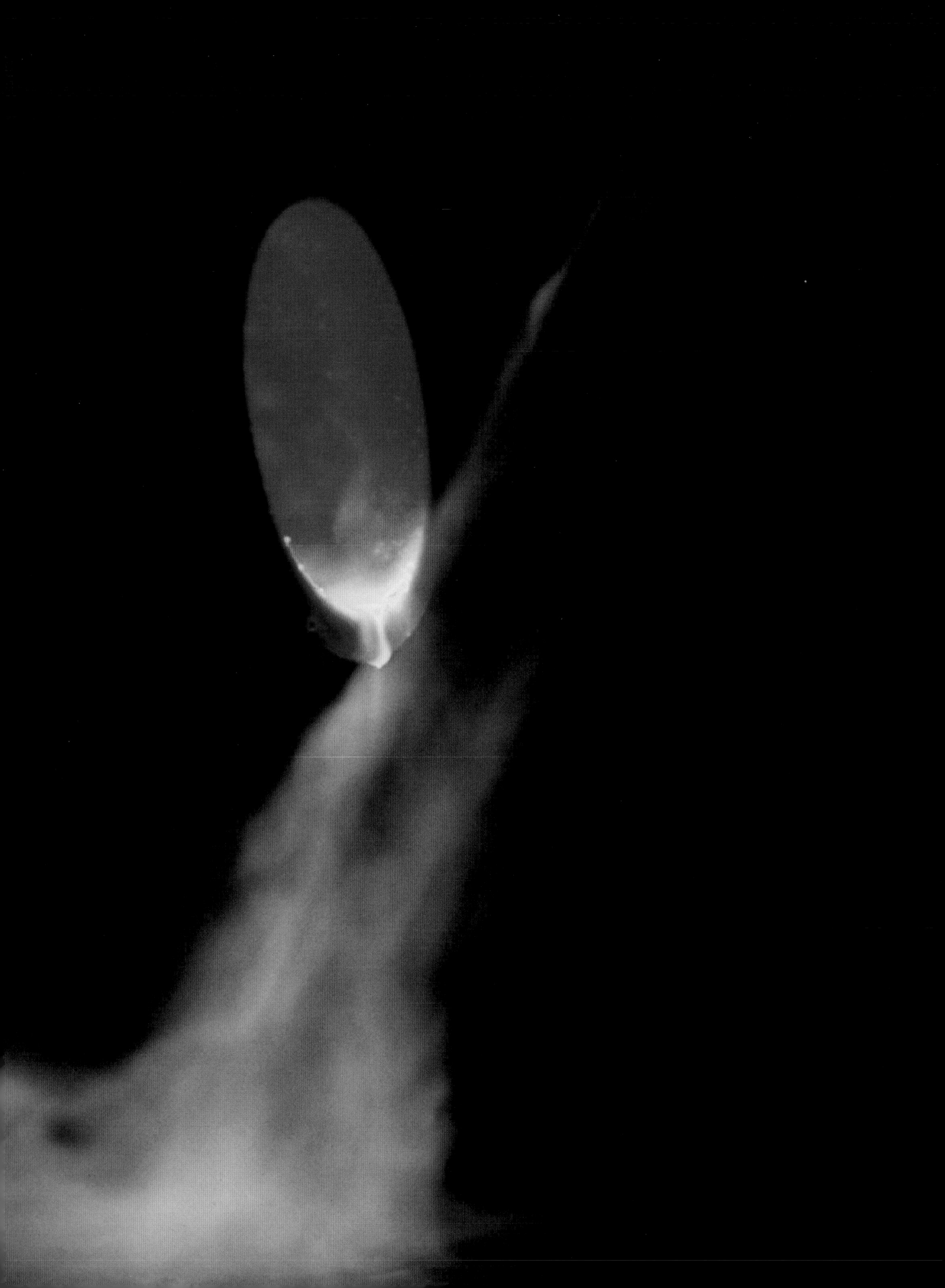

# Alloyed Forces

## ORDERING GOLD AND PLATINUM ALLOYS

BY DANIEL BALLARD, STEWART GRICE, AND JURGEN J. MAERZ

**Editor's Note:** When it comes to casting success, much depends on the metal itself. Over the years, *AJM* has presented numerous articles on how to best select alloys, including the following three:

• "The Golden Touch" (October 2000 *AJM*), in which Daniel Ballard focused on how to best match the choice of gold alloy with the intended manufacturing method.

• "Look on the White Side" (March 2002 *AJM*), an investigation by Stewart Grice into the properties of the common nickel-free white gold alloys.

• "Alloyed Forces" (March 2000 *AJM*), in which Jurgen J. Maerz described the characteristics of platinum alloy systems.

All three have been excerpted for this chapter, focusing on which alloys are best suitable for casting.

# Ordering Gold Casting Alloys

By Daniel Ballard

When selecting your alloy, the first thing to understand is that the manufacturing method—i.e., casting or fabricating—will affect the mix. Although you can ask for a multipurpose alloy, most of the time you'll be better off choosing a gold that's specifically formulated for one method or the other.

The most important difference between casting and fabricating gold is what happens to the alloy after it's melted. If it's cast into a lost-wax flask, a reaction can occur when molten metal meets hot gypsum investment. The common view is that overly heated metal and investment form a corrosive gas, such as sulfur dioxide, which could then affect the surface of the mold and, consequently, the metal.

To overcome such potential hazards, you'll most likely want an alloy that includes a deoxidizer, such as silicon or zinc. A deoxidizer not only promotes better filling but also works as a brightener: The trees break out cleaner, polishing goes much easier and faster, and such practices as bombing or electrostripping (both of which involve cyanide) can be avoided.

In my experience, a deoxidizer is most helpful in 14k or lower casting golds. With 18k alloys, its use depends on the melting temperature and the timing of the melt; in general, 18k seems to have less room for error in quench timing than the lower karats. Some of my customers swear by silicon in 18k casting; others swear at it. The subtlest differences in technique make this an individual issue, and one that all shops should be aware of.

You'll particularly want a "deox" alloy with an open-air melting method—such as resistance melting (which uses electricity and a high-temperature wire coil)—rather than one in which the molten gold is protected by a torch flame, kept under a cover gas, or, best of all, enclosed in an air-free environment. Your equipment will also determine the range of alloys you'll be able to accommodate. If you do use resistance melting, you'll want to stay away from such high-temperature golds as palladium white; 18k palladium melts at 2,000°F (1,093°C), and 14k palladium melts at 1,968°F (1,075°C). While some resistance machines can handle such temperatures, it will be a struggle if your equipment is old or worn out; you would be better off using a torch. Induction melters (which use radio frequency energy ranging from 6 to 500 kilocycles) will provide ample power to deal with any gold alloy. They also offer an air-free melting chamber, which means less deoxidizer will be needed.

Beware, though, when using silicon alloys, regardless of karat. Silicon can profoundly enlarge grain structure and greatly increase the possibility of brittleness and cracking. To help avoid this, make sure that you apportion the amount of silicon based on the job at hand. More silicon is needed in those cases where conventional polishing is most difficult or impossible (e.g., tennis bracelet links or stone-in-place setups where the stones can get in the way of polishing tools). Your refiner should be able to help you discover the proper level of silicon for your needs, as long as you provide detailed information about your tools and product.

In addition to a deoxidizer, casters should also ask their refiners about another additive that's recently come into use: the grain refiner. These refiners are actually trace amounts of a high-temperature element (often iridium) that is mixed into the alloy, and which serves as the basis for grain crystals. The best analogy I can give is the way in which raindrops form around bits of dust in the air; these dust particles are "nucleators" that provide a place for the raindrops to form. Grain refiners operate in much the same way as these particles. With their addition, the gold's grain structure becomes a network of very tiny crystals. This in turn provides for greater malleability and better surface finish, making polishing much easier.

As with silicon, though, grain refiners come with some big "ifs" attached to them. First, they need to be mixed thoroughly into the alloy, or large particles could cause inclusions. Second, the iridium and silicon could interact, with the silicon negating the beneficial effects of the grain refiner. The higher the level of silicon, the greater the chance of these problems occurring. Even the most advanced factory will use grain refiners with great care.

**Casting Colors and Karats.** When selecting your alloy, also be aware that not all colors and karats cast alike. Overall, alloys that are 18k or lower cast easier than those containing higher concentrations of pure gold, which tend to have larger grain sizes. These larger grain sizes could, in turn, make the alloy a little less strong and ductile.

Obviously, as you mix more alloying elements into your gold, the greater the impact those elements will have—and you best be aware of the consequences. Yellow gold, for instance, is less problematic than other gold alloys, regardless of karat. However, some alloys raise specific issues. Green golds, for example, can cast well, but they have a very

high degree of silver—80 to 92 percent, excluding the gold content. Consequently, they have a large grain structure. For best results, the tree should be quenched soon after the red color leaves the button.

With rose golds, you will have little trouble casting 10k and 14k alloys (although they could crack if quenched at too high a temperature). However, you might have difficulty working with 18k rose gold—in fact, you very likely *will* have difficulty. Because of the high copper content in the rose golds, the cast metal can become hard if it's allowed to air cool. And with 18k rose, I've seen jewelry pieces exhibiting such brittleness that, in some extreme cases, they've actually broken in transit to a customer (true story!).

Success with rose gold depends a great deal on when the metal is quenched (which should take place within a few minutes after casting). The difficulty with this lies in accurately measuring the temperature of the entire tree, which may not cool off at the same rate overall. The usual compromise I recommend is to add some silver to the rose gold mix, thus ensuring a decent grain structure while at the same time maintaining a nice rose color. If you use a non-silicon alloy with no less than 10 percent silver (excluding the gold content), problems should decrease significantly.

As for casting white gold, you must take into account the amount of nickel or palladium contained in the alloy. Most casters prefer the nickel alloy, which can provide a bright white color, is very hard, and is much cheaper than palladium. However, nickel white golds always pose a choice: To have castings that are either very white and very hard, or less white but more malleable for bending and setting. (Many manufacturers often plate their white gold with rhodium, which is a great way to increase brightness.) For casting, I recommend an alloy with 15 to 25 percent nickel (excluding the gold content) to achieve acceptable color and finish.

You might use the softer palladium alloy if the casting will have bead setting or some other detailed benchwork performed on it, or if you have hypoallergenic or regulatory concerns. [Editor's note: The European Union Nickel Directive requires that products used in piercing contain less than 0.05 percent nickel, and that jewelry that comes into direct and prolonged contact with the skin release no more than 0.5 micrograms of nickel per square centimeter per week.] If you're working with 10k or 14k, use 20 to 25 percent palladium, not including the gold content. Again, you may need to rhodium plate, depending on which other alloys, such as copper, are present. When you work in 18k, for best results use an alloy that is about 50 percent palladium, again excluding the gold.

## Using Nickel-Free White Golds

By Stewart Grice

Nickel-free white gold alloys were originally developed in the 1920s, with palladium the primary bleaching agent. For the most part, this is still the case today, particularly in 14k and 18k gold. Palladium has very good corrosion and tarnish resistance, and it mixes well with gold, offering almost complete homogenization throughout the range of gold-palladium compositions. It also yields alloys with excellent mechanical properties—superior in many respects to the nickel-whites.

But palladium does have one significant limiting factor: cost. Palladium white gold alloys can be quite expensive. As a result, the alloy you use often depends on the karat range you're working in and the type of jewelry you're producing.

For those who find the price prohibitive, there are other elements that will bleach gold to varying lesser degrees, but most have undesirable properties. Poor corrosion and tarnish resistance, high degrees of segregation, embrittlement, and poor mechanical deformation properties are often experienced with these palladium alternatives.

Following are the more common nickel-free white gold options in 10k, 14k, and 18k, with their benefits and drawbacks for casters.

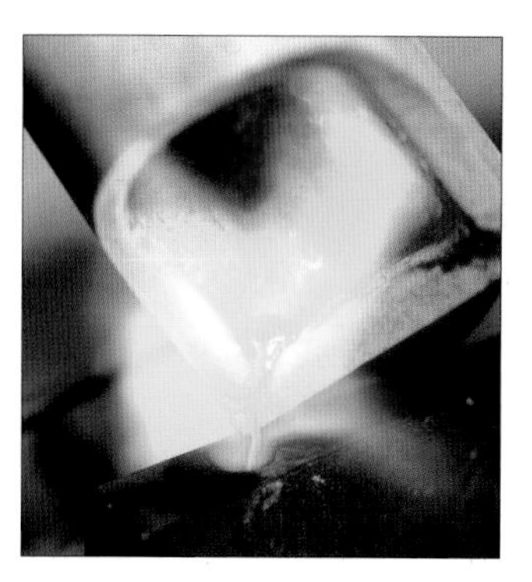

**10k nickel-free.** Most 10k nickel-free white golds on the market rely on high silver and zinc levels to bleach the gold. Some may contain a small amount of palladium—typically up to 5 percent—but cost is usually the limiting factor here. Most 10k nickel-free alloys also require rhodium plating as a final finishing operation to get the optimum white color. They have low hardness values when annealed, typically 95 to 115 HV. (Unlike copper, silver and zinc are poor hardeners of gold.)

These alloys do have a tendency to form dross during investment casting, a problem of zinc oxidation that is often encountered with high silver alloys. This can lead to crucible degradation, problems with pyrometric temperature measurement, and zinc oxide inclusions in the cast metal.

Oxidized zinc can also make the liquid metal sluggish, which may result in incomplete fills and metal failure, particularly in small sections like prongs. In addition, the zinc may present a problem if you are using a pressure-over-vacuum casting system. Most of these systems evacuate the top chamber prior to back-fill, and volatilized zinc is not good for your vacuum system.

**14k low-palladium alloy.** The majority of commercially available 14k nickel-free white gold alloys use palladium as the primary bleaching agent. The less expensive of the 14ks have a low palladium content; they are usually alloyed with silver, copper, and zinc. Silver is an important secondary bleacher: With the higher gold content it is not very efficient in a primary role, since it tends to impart a green hue.

Low-palladium alloys contain up to 10 percent palladium and the resulting color is not truly white, but what is often referred to as "straw white" or "cream white." (Rhodium plating offers the best results, but the alloy color will be noticeable when the plating wears off.) The hardness values of these alloys run from 95 to about 120 HV.

The low-palladium 14k alloys pose no significant challenges to investment casting. Adding palladium will always increase the melting range of a karat gold alloy, but the typical melting ranges for low-palladium alloys are 1,830°F to 2,010°F (1,000°C to 1,100°C). They should not present the caster with any major problems.

**14k high-palladium alloy.** The second sub-family of 14k nickel-free alloys is the high-palladium alloys. The big advantage here is color: These alloys are in excess of 10 percent palladium, and as the level increases, the alloy gets "whiter." The hardness values also increase, running from about 140 to 170 HV.

Unfortunately, nothing is free. The downside to these alloys is cost as well as greater age- and work-hardening characteristics, which will require additional steps in the production process. Higher melting temperatures also require special considerations in investment casting. As you add more palladium, the melting range moves even higher, typically to between 2,010°F and 2,190°F (1,100°C to 1,200°C), and possibly even higher.

This high melting temperature presents potential problems for the investment caster. More superheat (the difference between the liquidus temperature of the alloy and the temperature at which you cast it) may be required to prevent premature freeze-off and allow complete fills. This means that the casting equipment used for high-palladium white golds must be capable of reaching the required superheat for these alloys, which may be as high as 300°F (150°C) over the liquidus point of the alloy. In addition, the temperature measurement system must be capable of reading a temperature this high. (A type S thermocouple, for example, may be required.)

Different sprue and gate configurations may also be necessary: Those that work on a particular design in 14k yellow may be inadequate for a 14k high-palladium white alloy. Sprues may need to be thickened and shortened to prevent premature freeze-off and guarantee progressive solidification; multiple sprues and gates may also be required.

The higher casting temperature also creates potential problems with sulfate investments. When cast, the hot metal decomposes the standard sulfate investment to generate sulfur dioxide gas. This gas "pushes" the liquid metal away from the mold wall and leads to a rough surface texture and a heavier oxide layer, which means you have to do more finishing work. This problem can be avoided with the use of a high-temperature investment.

For both spruing and investing, individual specifications will vary. The appropriate method will be determined through experience.

**18k low-palladium alloy.** Melting temperatures are in the range of 1,740°F to 1,920°F (950°C to 1,050°C), so there are no surprises for the investment caster. Hardness values range from about 120 to 140 HV.

**18k high-palladium alloy.** Melting temperatures are in the range of 1,900°F to 2,000°F (1,035°C to 1,100°C) at the lower end, and 2,150°F to 2,280°F (1,175°C to 1,250°C) at the higher end, depending on composition. Hardness values range from about 95 to 130 HV. (For information on casting, see instructions for 14k high-palladium casting.)

## Platinum Casting Alloys

by Jurgen J. Maerz

The world of platinum is not one into which you can step without ample information: There are many variations among the platinum alloy systems, each with a specific purpose. The following guide offers insight into the casting capabilities of these alloys. As always, for best results, make sure you take the time to understand your alloy and choose well.

**Platinum/iridium systems.** To many jewelers, the platinum 900/iridium100 alloy is a dream come true. Ductile and malleable, it can be welded with the torch, cast, machined, and stamped. As it does not markedly oxidize, no flux or pickle is required. It has a Vickers hardness of 110, a density of 21.5, and a melting range of 3,272°F (1,800°C) liquidus to 3,236°F (1,780°C) solidus.

However, while 900/100 is preferred, a platinum content of 950 parts per thousand is required in the U.S. and many foreign countries in order to mark jewelry "Platinum." To meet this standard, casters in the U.S. have started using platinum 950/iridium 50, which is actually less expensive but works best as a fabricating alloy. This system has a Vickers hardness of about 80, however, which will lead to soft castings—rings will scratch rapidly and lose their shine, for example

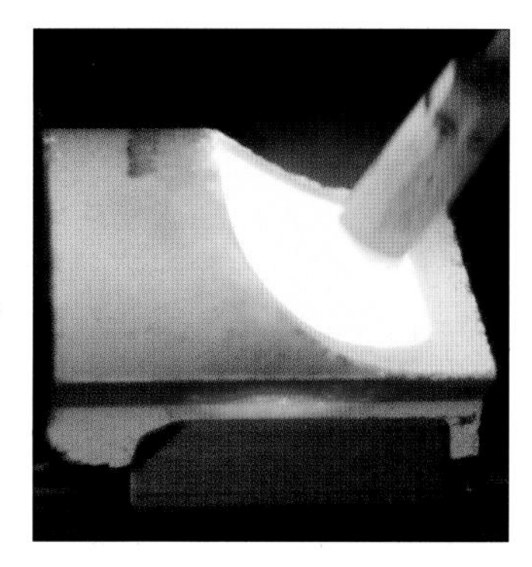

**Platinum/cobalt systems.** Combined with platinum, cobalt acts as a grain refiner and is used solely for casting. In fact, platinum 950/cobalt 50 is one of the most superior casting alloys on the market today. It is being used in the U.S., Europe, and Hong Kong. Its good flow characteristics make it possible to fill even the finest details in castings. It is important to note, though, that this alloy is slightly ferromagnetic and oxidizes under certain conditions. The 950/50 alloy has a melting range of 3,218°F (1,770°C) liquidus to 3,056°F (1,680°C) solidus.

Because platinum/cobalt tends to oxidize in atmosphere, casting with it should be done with atmosphere control or vacuum. For the same reason, while casting the alloy with a torch setup is possible, it is somewhat difficult with any other fuel but hydrogen/oxygen. The same fuel should be used when welding the alloy, which can also be successfully laser-welded.

**Platinum/palladium systems.** Palladium systems are all soft, ranging from 60 to 80 Vickers in hardness, and tend to form cavities in the casting. Jewelry made from platinum/palladium alloys requires extensive burnishing to work-harden the surface and ensure a fine polish. Furthermore, these alloys have a grayish color; in Japan, many products cast with this alloy are subsequently rhodium-plated.

**Platinum/gold systems.** If you are looking for a general-purpose alloy, a platinum 950/gold 50 system will do very well. It has a Vickers hardness of 90 and is a fine casting alloy that, because gold won't oxidize, does not require a special atmosphere. However, as is typical with this alloy system, the casting needs to be quenched immediately to prevent hardening and brittleness.

Another fine all-purpose alloy is platinum 900/gold 100, which has a Vickers hardness of 135 and a density of 21.3. Used for fabrication and casting, it has a melting range of 3,191°F (1,755°C) liquidus to 3,110°F (1,710°C) solidus and should be cast at about 3,290°F (1,810°C). It is popular in Japan, Europe, and South Africa.

**Platinum 950/ruthenium.** Platinum/ruthenium can be difficult to cast because it is not very fluid, but experienced casters using induction melting techniques can achieve satisfactory results. Torch melting is not recommended, though, since ruthenium oxide fumes are toxic.

WARNING
NIOSH
WARNING

# Casting Safety

## MINIMIZING RISKS, FROM BURNS TO TOXIC FUMES

BY CHARLES LEWTON-BRAIN

Jewelers cast metals in various ways, from using molds made of dried sand to employing high-tech vacuum or centrifugal systems. Yet, as different as they may be, all of these processes share one important characteristic:

Each can be dangerous without the proper safety precautions.

Think about it: Casting involves heat, with the attendant dangers of burns and damaging infrared and ultraviolet light. In lost-wax investment casting, the burnout process (which normally occurs in kilns) produces fumes that range from mildly to extremely toxic, and which in many cases are fairly polluting. Many alloys cast in the jewelry workshop—gold, silver, platinum, and, less commonly, brass, tin, and zinc—may contain other metals that produce hazardous fumes, such as lead, cadmium, beryllium, arsenic, and antimony. Nickel, too, is considered a suspect metal now, and is often found in casting alloys. Manganese and chromium, often found in re-used metals, should also be avoided.

It's not just when the torch or casting machine is fired up that problems can occur, of course. The preparations for casting include solvent exposure, tool use, and mold making (those scalpels are sharp!), and post-casting steps include devesting and finishing operations. But since covering every hazard in the entire process would take an entire book, the following sections focus on the dangers present in the investing, burnout, and casting processes, as well as general related hazards. This is a very brief introduction to safety issues, and it cannot hope to address the whole gamut of problems. But it can serve to introduce some important principles.

## General Casting Hazards

General casting hazards can be divided into four categories:

**CHEMICAL:** The materials, waxes, metals, and investments we work with are all chemicals, and they can affect the body. The fumes from waxes during burnout are dangerous, as are metal fumes. Toxic metals such as cadmium can also be found in older metal that is remelted. And cristobalite from investment powders can lead to silicosis.

These chemicals can be absorbed into the body through inhalation, ingestion, and skin contact. As a matter of principle, don't eat, drink, smoke, bite your nails, or apply makeup in the workshop. Smoking, besides being bad for you in all the ways you ought to know about by now, seems to react synergistically with many chemicals and dusts jewelers have been exposed to. In some cases, it dramatically multiplies the risk of damage.

Remember, too, that heat not only causes chemical reactions to accelerate, but also sometimes starts them—and heat is definitely part of the casting process.

**PHYSICAL:** These hazards include burns, cuts, crushings, electrocution, accidents, and radiation (ultraviolet and infrared light can damage the eyes). Noise doesn't pose much of a hazard for casters, although torches and other loud shop noises could cause hearing loss.

**ERGONOMIC:** These problems include working heights that are too tall or too short, too few breaks in work, and repetitive movements. Make sure that you do not have to be positioned awkwardly at any point during the casting procedure or process. If you're casting as part of a production system, be sure to examine the ergonomic issues quite carefully.

**FIRE:** When working with gases, torches, hot metals, and electric and gas kilns, you always run the risk of a fire or an explosion. Care and proper maintenance are especially critical when you use a hand torch: Keep all flammable materials and liquids away from any source of heat or spark, and make sure you have fire extinguishers on hand (as well as a fire plan).

In these four categories, the main dangers are silicosis caused by breathing in cristobalite (the primary ingredient in investment powders); dermatitis from repeated skin contact with investment, debubblizers, and fluxes; lung damage or illness caused by breathing in metal fumes; eye damage caused by not using the right kind of eye protection against ultraviolet or infrared light, or by flying chunks of metal or investment; and burns and accidents with the electrical, gas, and mechanical equipment used for casting.

The next sections list the hazards present in each procedure, as well as some suggestions to help minimize them. For the most part, I consider the casting process to be very safe, but you should understand that possibilities for harm do exist—and I have by no means covered all the potential dangers. Safety is your responsibility, and it is up to you to consider safety in your own particular circumstances.

## Investing

**CHEMICAL:** Potentially hazardous chemicals include alkalis in investment, which can foster dermatitis with skin contact, and detergents and defatting agents in debubblizers.

**PHYSICAL:** Common hazards include electrocution from electrical short circuits, accidents due to a tripping hazard or clutter in the workshop, and splashed investment in the eyes. Most critically, you must beware of inhaling cristobalite. The main investments used by jewelers are up to 60 percent cristobalite, a form of crystalline silica that is much more dangerous than ordinary silica and can lead to a serious risk of silicosis. It is also a possible carcinogen.

**ERGONOMIC:** Ergonomic injuries generally occur only in real production situations, though working height and stance should be considered in any job.

**FIRE:** Fire hazards stem mainly from electrical wiring problems. Always make sure the fusing is correct for the demands on your electrical service. Check and maintain wiring on any electrical device.

**EXPOSURE ROUTES:** Eyes, inhalation (of silica), skin.

**SAFETY PRECAUTIONS:** Use local ventilation to suck the airborne investment particles away from where you are mixing and measuring. Wear a mask when investing, although a mask is no substitute for good local ventilation. Try to control dust by mopping floors and sponging surfaces with water. Use latex or vinyl gloves to protect your skin when touching the investment.

## Burnout

**CHEMICAL:** Potential hazards include wax and plastic fumes from burnout, and fumes from any gas heat source. If you use polyurethane or Styrofoam instead of wax for the positive mold, burnout can produce hydrogen cyanide and other toxic gases (McCann, *Health Hazards*, pages 51-52). Styrofoam also produces carbon monoxide when burned (Monona Rossol, *The Artist's Complete Health and Safety Guide,* page 277). Using Styrofoam in this way is not recommended.

"When…waxes burn, they release many toxic and irritating compounds, including acrolein and formaldehyde. Acrolein is an exceedingly potent lung irritant, formaldehyde is a sensitizer and suspect carcinogen," writes Rossol. Burning organic chemicals like rosin, petroleum jelly, or mineral oil, which may be added to wax, "will release carbon monoxide and other toxic decomposition products" (Rossol, page 277).

**PHYSICAL:** Hazards in this category include electrocution from electrical short circuits (this happened to me once in Germany when I reached into a kiln); accidents due to tripping hazards or clutter in the workshop; burns while checking the progress of burnout; and irritation from wax fumes. There is also the rare possibility of investment popping and hitting you while you're inspecting it.

**ERGONOMIC:** Burnout does not involve many ergonomic issues, except for working heights, confined spaces, and access to equipment.

**FIRE:** Electrical and gas kilns are the most obvious fire hazards. If the kiln is opened before the wax residue is gone, there may be a flare-up, and billowing flames can erupt. Just close the door to put out the fire.

**SAFETY PRECAUTIONS:** You should have eye protection when near the flasks, and wear protective gloves and clothing when handling them. To move the flasks during burnout, use gripping tongs that do not slip.

**SUBSTITUTION OPTIONS TO REDUCE RISK:** Autoclaving can take care of some of the potential problems by removing the wax in a liquid form, which means you don't convert the waxes into toxic fumes. Avoid Styrofoam and consider using waxes such as beeswax, which produces fewer toxic fumes than some of the plasticized waxes. And, of course, you can always outsource.

## Casting

**CHEMICAL:** Potentially hazardous chemicals include borax, boric acid, and possibly other casting fluxes, such as ammonium chloride or sodium chloride. Zinc is also a concern, if you use it to de-gas a melt. Some alloys may contain hazardous metals, such as cadmium, beryllium, arsenic, and antimony. Metal fume fever causes flu-like symptoms, including fever, chills, and aches, usually two to six hours after exposure (Rossol, page 133). The symptoms last for 24 to 36 hours (McCann, *Artist Beware!*, page 425). Unless the metals are toxic, such as cadmium, those exposed seem to recover with no long-term effects.

Be particularly wary of using non-precious scrap metal for casting, as it may contain metals that will produce toxic fumes when melted. Scrap and reclaimed "mystery" metals may also be coated with lead- or cadmium-based paints, which produce poisonous gases when heated.

**PHYSICAL:** Like the other procedures, casting poses electrical hazards: If you are using an induction melting crucible or an electric kiln, then electrocution is a risk, as is fire. Other considerations: accidents due to tripping hazards or clutter; injuries from mechanical parts cracking or smashing into a hand that is incorrectly positioned; and burns suffered while removing and preparing flasks for casting.

Accidents can also occur if your casting machine is not balanced correctly, or if you have not maintained and checked it properly before use. (Such carelessness can result in a worker's broken arm or in red-hot metal globs being splattered across the room.) Other dangers include the risk of an explosion if you're using gas torches, and the breathing in of investment dust, which is particularly present during centrifugal casting and can cause lung disease.

**ERGONOMIC:** In casting, such issues primarily involve working heights, movement positions, and a setup that doesn't require twisting and turning your body. In a production situation, you would have to examine the workspace and the job very particularly to ensure that ergonomic issues are addressed.

**FIRE:** When using electrical equipment as the heat source for melting, all issues of electrical safety must be addressed. These include current draw, fusing, wire conditions, maintenance, and the possibility of electrocution. When using gas kilns or torches, you must pay attention to such issues as checking for leaks, proper storage, and the proper use of equipment, such as flashback arrestors.

**EXPOSURE ROUTES:** Inhalation (wax and metal fumes, dusts of various kinds, fumes from fluxes, gas), skin (fluxes and other materials, burns), eyes (infrared and ultraviolet radiation, and possibly sodium flare).

**SAFETY PRECAUTIONS:** Use local ventilation. You are dealing with hot objects, so the kiln and casting spot must be in a segregated, fireproof area. All flammable materials and liquids must be kept away from heat. Fireproof walls should surround the kiln area. A burn kit and ice should be on hand in case someone is burned. Ensure that you have a fire plan available. The casting area should be damp-wiped frequently to remove cristobalite dust.

The proper maintenance of all equipment, including the torch systems, is critical. You'll also want to be sure you have good ventilation, as well as eye protection from both particles and radiation (infrared, ultraviolet, sodium flare). Eye protection is particularly important when you're melting with an oxy-acetylene or oxy-propane torch system, since these emit significant amounts of ultraviolet light.

Use heat-protective gloves, a leather apron, and steel-toed boots with molten-metal splash-proof covers over the lace area to protect against spills. Place a shield around the casting machine to keep metal splashes segregated: The old washtub around the centrifuge is a reasonable shield.

Be familiar with all materials used—the fluxes, the metals, and the chemicals involved—as well as their effects. Make sure that copper, for instance, does not contain any beryllium. If you're not familiar with the possible safety hazards of a material, ask your supplier for a Material Safety Data Sheet (MSDS).

**SUBSTITUTION OPTIONS TO REDUCE RISK:** Outsource. Switch to other types of manufacturing procedures, such as metal stamping, construction, milling, and computer-aided design and manufacturing. Substitute materials are possible. For instance, it's apparently possible to use a non-cristobalite or lower-cristobalite investment for some casting purposes. Or try to use waxes for burnout that do not contain multiple chemicals. Sand casting may substitute for investment casting.

**REMEMBER:** No matter the size of your operation, always set things up so you can't hurt yourself. Make safety a habit, and it won't seem like work. If you have an accident or a near miss, record it in an "accident book" and discuss it with employees, friends, or colleagues. Recording the event will help remind you to avoid doing the same thing again.

And to avoid unnecessary work, look for people who have solved the same kinds of safety problems you have experienced, and see if you can adapt some of their ideas and approaches for your own shop. There are role models all around you, if you just take the time to look—and make safety a part of your regular thought pattern.

**References:**

McCann, Michael. *Artist Beware!* Rev. ed. New York: Lyons and Burford, 1992.

McCann, Michael. *Health Hazards Manual for Artists.* 4th rev. ed. New York: Lyons and Burford, 1994.

Rossol, Monona. *The Artist's Complete Health and Safety Guide.* 2nd ed. New York: Allworth Press, 1994.

*This chapter was adapted from* The Jewelry Workshop Safety Report *(Brain Press, 1999) by Charles Lewton-Brain. The book is available from Brain Press Ltd., 1-403-263-3955.*

# A Matter of Maintenance

## CASTING MACHINES NEED RESPECT, TOO

BY ELAINE CORWIN

John arrived at work 20 minutes early. He felt pretty good about that, especially since he was feeling pretty bad overall: He and Annie, his girlfriend, had had another fight. She'd been kind of cool toward him ever since he forgot their one-year anniversary last month. Then last night she told him she needed time to think—"downtime," as she put it.

"Why can't things be like they were in the beginning?" he'd asked.

"John, we're not new anymore," she'd said. "It takes a little maintenance now." What did that mean? John shook his head. Women!

"Well, maybe I'll start casting early today, take my mind off of it," he said to himself. He walked over to the casting machine and flipped on the breaker. The machine gave a low groaning sound and the lights flickered dimly. It had been doing that a lot lately, John thought. He kept meaning to check it out, but he never seemed to have the time. Anyway, it usually stopped in a couple of minutes.

But not today.

"A little mechanical correction never hurt," he thought, and smacked the machine on its side. The machine sputtered, then a small envelope popped from the flask chamber. Surprised, John picked it up and opened it. A letter—and it was addressed to him!

"Dear John," it said. "We can't go on like this anymore." John turned the letter over. It wasn't signed, and it really didn't look like Annie's handwriting. Was this some kind of joke? He read on.

"You don't treat me well anymore, but you expect me to perform for you. Well, I just can't do it. I'm ready to blow a gasket, my filter is clogged, and you haven't changed my oil in months. I look like hell, and I suppose you think it's my fault—that I've let myself go!

"Remember the good times, John? When I was new? You were so good to me. You read my manual. You checked my oil level every week. And I produced for you—we really turned out some great castings together. I gave you the best months of my life."

This is ridiculous, John thought. A machine can't talk, much less write a letter.

"You were so proud of me then, John," the letter continued. "I remember when your friend Joel told you that his machines hadn't been down in more than 13 years. He said he just made it a habit to maintain them regularly. He always stocked spare parts and he kept them under lock and key, so he'd always know where they were. He inspected gaskets and o-rings every month, and he replaced any that looked worn. Joel even helped you make out a maintenance card for me. It was beautiful! It had lines so you could schedule our service dates. Where is that card now, John?"

"The card," John said. "I forgot about the service card." He checked the side of the machine. Sure enough, the card was still there, but it was covered now with dust and grime. It had only two service dates filled in. The last one was…

"Oh jeez, has it been that long?" Now he was ashamed.

"You've been taking me for granted, John, so I'm leaving you. I need some time to think—downtime."

"Downtime?!" John shouted. "Oh no, just like Annie. Wait! Please, don't quit on me now! I've got an oven full of flasks! I'll change, you'll see. Just give me another chance."

John spent the next two hours cleaning the machine. He checked all the gaskets, belts, and o-rings, then made a list of everything his machine would need to function properly. He got down on his hands and knees and vacuumed the filter—a job he didn't particularly like, but it made her so happy. He found the manual in the bottom of his desk drawer, then put in a locking cabinet to hold all the spare parts he was going to order.

When he was finally done, he filled in the date on the service card and wrote the date of the next maintenance on his calendar. Then he looked at his machine. It was beautiful. Things would be different now, he knew, and for the first time that day he smiled. He called his supplier to order all the new parts and spares.

After hanging up with his supplier, he picked up the phone again. This time he called the florist and ordered for Annie the biggest bouquet they had.

"What's the occasion?" the florist asked.

John smiled. "Maintenance."

# Cast Aside Your Troubles

## TROUBLESHOOTING COMMON PROBLEMS

BY BOB ROMANOFF

**Editor's Note:** A great many factors contribute to a successful casting. Rubber molds must be properly vented. Wax patterns must be defect-free and securely attached to the sprue. Investment must be adequately mixed to the right consistency. Flasks must be positioned correctly in the oven, and burned out at the correct temperatures and for the correct amounts of time. The metal must not be overheated or contain contaminants. And the list goes on. Make even one mistake at any step in the process, and you can wind up with not only imperfect castings, but also hours of troubleshooting.

However, even the best casters make mistakes. If you do end up needing to retrace your steps and determine where you went wrong, the following guide to some of the more common problems can keep the head-scratching to a minimum.

## Wax Injection

**Problem:** Rough surface.

**Cause 1:** Too much powder on the mold.

**Solution:** Use additional cloth layers in the powder bag to minimize the powder on the mold. Apply only to vents.

**Cause 2:** The wax temperature is too hot.

**Solution:** Keep wax temperature between 150°F to 160°F (65°C to 71°C), depending on the particular wax.

**Cause 3:** Using both powder and silicone spray at the same time on a rubber mold, resulting in a residue buildup.

**Solution:** Use a silicone spray only to remove very difficult undercut wax patterns. Fine powder should be used every time, but only sparingly on each half of the mold with the mold bent open, exposing the air vents. When the mold is allowed to flatten, blow off the excess powder from the surface of the design.

**Problem:** Incomplete fill.

**Cause:** Insufficient wax temperature, insufficient pressure, or insufficient air release in the mold.

**Solution:** Cut additional air releases wherever the piece does not fill in, or gradually increase the temperature and pressure.

**Problem:** Excess shrinkage or sink holes in heavy pieces, especially pieces with flat surfaces.

**Cause:** Shrinkage of the wax while it is cooling.

**Solution:** There are several ways to address wax shrinkage. In many cases, the best solution is to simply hold the mold against the wax injector until the wax is completely solidified, generally 15 to 20 seconds. You can reduce this time until the shrinkage occurs again. In addition, try injecting the wax at the coolest possible temperature and using a quicker drying wax.

**Problem:** Waxes with non-uniform weights.

**Cause:** Some wax injector operators hold the rubber molds with greater pressure than do others.

**Solution:** Use an automatic mold clamp that attaches to your wax injector. The clamp will keep the pressure on the molds uniform.

**Problem:** Bubbles in the wax.

**Cause:** Insufficient number of air releases in the rubber mold.

**Solution:** Cut additional air releases and properly apply powder in vents. [Editor's Note: See also "Finding Flaws," page 32, for additional advice on preventing wax defects.]

## Investing

**Problem:** Investment sets too quickly.

**Cause:** Water temperature is too high, the mixing machine beater is spinning too fast, or the batch of investment may be bad.

**Solution:** Use the correct water temperature suggested by the investment manufacturer—usually 70°F to 75°F (21°C to 24°C)—slow down the mixing machine, or change the batch of investment.

**Problem:** Water separation marks.

**Cause:** These marks, which look like sand running down a smooth surface, are caused by trying to rush the investing time cycle. The normal work time of most investments is 9 to 10 minutes. If you finish working with the investment 2 or more minutes before it begins to solidify (gloss off), the powder begins to settle to the bottom of the flask. As the powder settles, it leaves marks on the waxes that later show up on the castings.

**Solution:** Increase your mixing time so the flasks are being removed from the vacuum chamber with less than 1 minute before gloss off.

[Editor's Note: See also "Investment Advice," page 44, for more about avoiding investing problems.]

## Burnout

**Problem:** After casting, investment color is gray, brown, or black.

**Cause:** These colors are usually the ash residue that remains in the investment from wax that has not been burnt out completely. Insufficient burnout occurs when the temperature in the oven does not reach 1,350°F (732°C) or has not remained at that temperature for a long enough period of time. After casting, the investment should be chalk white throughout the flask.

**Solution:** Correct the temperature of your burnout oven, use a longer burnout cycle, or use a steam dewaxing oven to remove the wax from the flasks before the burnout cycle. (Steam dewaxing is necessary for low-temperature stone-in-place casting.)

**Problem:** Flasks in different parts of the burnout oven are at different temperatures.

**Causes:** In gas ovens, variable temperatures are caused by the following:

- Badly deteriorated bricks. Crumbling bricks reduce the

insulating effectiveness of the walls.

• Broken hearth plate. "Hot spots" will occur at the cracks.

• Blocked gas tubes. The burner flame will not be uniform if the individual gas tubes of the burner assembly are blocked by powder from the bricks or flasks.

• Door doesn't close tightly or has gaps. Gaps in the door cool the front of the oven, so the flasks in the front will be cooler than those in the middle or back.

• The exhaust system isn't working properly or isn't strong enough. Poor exhaust results in insufficient air circulation inside the oven, causing non-uniform temperatures.

• Exhaust is too strong. If the exhaust is too strong, the oven may never reach the temperature required for proper burnout.

**Note:** In electric ovens there is little or no air circulation, so flasks must rotate inside the oven in order for each flask to reach a uniform temperature.

## Casting

**Problem:** Fins on the castings.

**Cause:** Rough handling of the flasks during the first hour after investing.

**Solution:** Handle the flasks very carefully before they are placed into the burnout oven. Flasks should sit on a vibration-free table.

Fins can also be caused by the following:

• Poor quality or bad investment.

• Too much water mixed with the investment. Solve this problem by following the investment manufacturer's instructions for the correct water-powder ratio very carefully.

• Temperature in the burnout oven above 1,400°F (760°C). If you think your burnout temperature is the source of the problem, recalibrate the oven temperature.

• Temperature increased too rapidly during burnout cycle. Use the following burnout cycle:

1 hour holding at 250°F (121°C);

1 hour climbing up to 600°F (316°C);

1 hour climbing up to 1,000°F (538°C);

1 hour climbing up to 1,350°F (732°C);

2 to 4 hours at 1,350°F (732°C) (large flasks, 4 hours);

1 hour gradually cooling down to 1,150°F (621°C);

1 hour gradually cooling down to 1,000°F (538°C) or your desired casting temperature.

**Problem:** Rough surfaces on the casting.

There are four common causes of rough surfaces.

**Cause 1:** Burnout temperature rises too rapidly during the first hour, causing the wax to boil in the investment before it has the chance to drain.

**Solution:** Hold the temperature at 250°F during the first hour of your burnout cycle.

**Cause 2:** Branches of the wax tree are pointing straight out or at a downward angle. These positions prevent the wax from draining even during a correct burnout cycle. The trapped wax boils in the investment. If this is the source of your problem, the roughest part of the casting will be the part that is at the low end of the tree branch.

**Solution:** Keep the branches of your wax tree pointing upward, usually at a 45° angle.

**Cause 3:** Your wax wash is contaminated or you invested the flasks before the wax wash was completely dry.

**Solution:** Use new wax wash and wait for flasks to dry completely.

**Cause 4:** Wax patterns with excess powder on their surfaces.

**Solution:** Use less powder, or clean off excess powder with wax wash.

**Problem:** Distorted castings.

Castings can become distorted due to several factors. Three causes are most common.

**Cause 1:** Wax patterns that were bent before they were set on the tree.

**Solution:** Inspect wax patterns carefully through an illuminated magnifying lamp before setting them on the tree. For best results, use a dark, opaque wax that allows defects to be easily seen.

**Cause 2:** The wax is too soft and bends when the investment is poured into the flask.

**Solution:** Use a firmer wax and always pour investment between the wax and the inside wall of the flask.

**Cause 3:** Castings are bent when they are cut off the tree or when the sprue stump is cut off. Use extra care.

**Problem:** Shrinkage porosity (dark cracks or criss-cross surface patterns and rough, pitted surfaces).

**Cause:** Uneven solidification of the piece.

**Solution:** Provide larger sprues or additional sprues so uniform solidification can take place. Solidification should occur from thin to progressively thicker sprues, then to the sprue rod, and finally to the button.

The ideal sprue is 1 to 1.5 times thicker than the heaviest part of the casting, and it feeds directly into that section. Obviously, your design may not always allow you to use the "ideal sprue," but keeping its requirements in mind will help provide the progressive solidification necessary for minimizing porosity.

In addition, flask and metal temperatures must be as cool as possible—just hot enough to fill the entire tree. When flask and metal temperatures are too high, the cooling rate is erratic, which can cause shrinkage porosity.

**Problem:** Gas porosity. This shows up as a group of small holes, usually in shanks and shoulders of rings, but it can appear almost anywhere. Severe gas porosity will actually look like squeezed air bubbles.

**Cause:** Several conditions may cause gas porosity:

• Severe overheating of metal.

• Wrong burnout time and temperatures.

• Weak oven exhaust system.

• The exhaust is too strong and blows contaminants around the inside of the oven and into the flasks.

• Flames are too high in a gas burnout oven. Adjust the flame so it appears blue with a yellow tip and is just below the surface of the hearth plate.

• Flasks overhang the hearth and are in contact with flames. Be careful not to crowd the oven. Some space is needed around each flask for good circulation, which promotes uniform flask temperature.

• Remelting with investment still on metal. Carefully examine metal for investment, oxides, etc., before remelting. It is best to remelt metal at the lowest possible temperature and form shot prior to casting. Also, never use less than 50 percent fresh metal.

If your shot looks like corn flakes, the water temperature is too cold. If it bunches together like scrambled eggs, the water temperature is too hot. When pouring shot, the water in the tank should be circulating and at a temperature of 100°F to 130°F (38°C to 54°C).

• Contaminants in metal or crucibles. Use only shot or fresh metal for casting. Making shot from re-melted metals will usually separate most contaminants from the metal.

**Problem:** Incomplete castings.

**Cause 1:** The flask or metal temperature is not hot enough for the metal to flow into all the details of the casting.

**Solution:** Gradually increase the temperature of the metal or the flask until all the pieces fill in completely.

**Cause 2:** Insufficient suction in vacuum casting machines.

**Solution:** The vacuum pump may not be working properly, or the filter or vacuum hoses may be clogged. The flask gaskets of the casting machine may also need to be replaced. Be sure to change the oil in the vacuum pump frequently.

**Problem:** Shiny "moon craters" in the castings.

**Cause:** An excessive amount of boric acid flux added to the melt.

**Solution:** Use less flux or no flux at all if metal is melted under protective gas or in a vacuum.

**Problem:** Black spots.

**Cause:** Graphite flaking off the crucibles and being drawn in with the molten metal as the metal enters the flask.

**Solution:** Use a better quality graphite crucible and glaze the inside wall of the crucible with boric acid. This glaze will prevent the graphite from flaking off, which will also lengthen the life of the crucible. Be sure to pour out all the excess boric acid to prevent boric acid inclusions.

**Problem:** White spots.

**Cause:** Particles of investment have flowed in with the metal.

**Solution:** After removing the flask from the oven, lightly tap it with the opening face down. This allows any loose particles of investment to fall out.

**Problem:** "Balls" on the casting.

**Cause:** These lumps are usually the result of insufficient vacuum during investing. Insufficient vacuum can be caused by a weak vacuum pump, a leak in the plumbing, or insufficient vacuuming time.

**Solution:** Continue vacuuming the investment for 45 seconds after the investment rises and then falls. During vacuuming the investment should be "boiling" and actually splattering against the bell jar.

# All to Pieces

## FROM SMALL ERRORS, BIG PROBLEMS GROW

BY TIMO J. SANTALA

After a month of setting up a casting facility on the other side of the country, Raul was happy to be back home—until he arrived at work and found a mess in the casting department. The trees from that morning's cast had one problem after another: porosity, finning, flashing. He immediately sought out the company's new casting manager—his old student, Richard.

Raul found him outside the wax burnout department, a stack of chart recordings in one hand and a frown on his face. "Richard!" he yelled. "What's been going on here? The trees look horrible!"

"I know," Richard said. "When you left, everything was casting fine, but now we're getting nothing but bad results. I just can't put my finger on the problem. I thought I could do it all on my own."

Raul sighed. Obviously, his days as a teacher were not over; Richard would once again have to learn the hard way. "Let's see what happened," Raul said. "I want to start at the beginning and work with every line manager until we figure out the cause of this problem."

They started in the wax department. "Richard," Raul said, pointing to a recently sprued tree, "how do these waxes look to you?"

"They look pretty good—not perfect, but pretty good," Richard hedged.

"You're right. But look, there's too much parting powder here, and the pieces are a little close together. Also, we should try to clean up the gate attachment to the tree by smoothing it a bit more. But you're right, it is pretty good."

They moved to the investment room. "How are we doing in here?" Raul asked.

"I think we're under control," Richard said. "The guys have been following the standard operating procedures all month and have noted any anomalies."

"Good. I see only one minor thing," Raul said. "Look at the rubber bases; some of them have a little buildup of wax. We should try and clean that up if we can. But you're right again, we look pretty good here."

Two areas down and no major problems had been found, but Richard didn't know whether to feel better or worse. After all, they still hadn't found what had caused the horrible castings.

"All right, let's go into the casting room," Raul said.

Richard followed nervously as Raul made his inspection from one end of the room to the other. "Once again, not bad," Raul commented. "I see a few things I'd change, but overall not bad."

"What things would you adjust?" Richard asked.

"Well, a few pieces of scrap still had tiny amounts of investment on them, the buttons on some of the trees were too small, the caster moved one or two of the flasks a little too soon, and a few of the flasks seem to be seated imperfectly, but the vacuum was still good. Like I said, nothing major. Let's look at breakout."

Raul spent a few minutes in the breakout area before walking out. "Once again, everything looks decent," he said.

"Well, what can it be then?" Richard asked, puzzled. "We've looked at every area. Something must be wrong."

"Oh, there is," Raul said. "Richard, things are out of control."

"But... But... You said...!"

"I know what I said at each and every step. I said things looked all right, BUT... That 'but' is the problem. Casting is a series of steps, with each step dependent on the one before: Waxing, treeing, investing, burnout, and casting. If we have a 90 percent success rate at each step, at the end of the run, we get 60 percent of the castings without a problem. Now let's say that half the time, one problem overlaps another, so that we have only one casting affected rather than two. If we overlap half, still only 80 percent of the castings come out good."

"I see. So we need to tighten up the little things," Richard said.

"That's right. We can't control casting 100 percent, but I bet if we take care of all these little things, tomorrow's castings will look better than today's. And it will be a lot easier for us to find problems in the future if we're sure about all these details. Right, Richard?"

"Right, Raul. I'm glad to know I can always count on you. Speaking of details, let me show you some chart recordings..."

Raul shook his head. A mentor's work is never done.

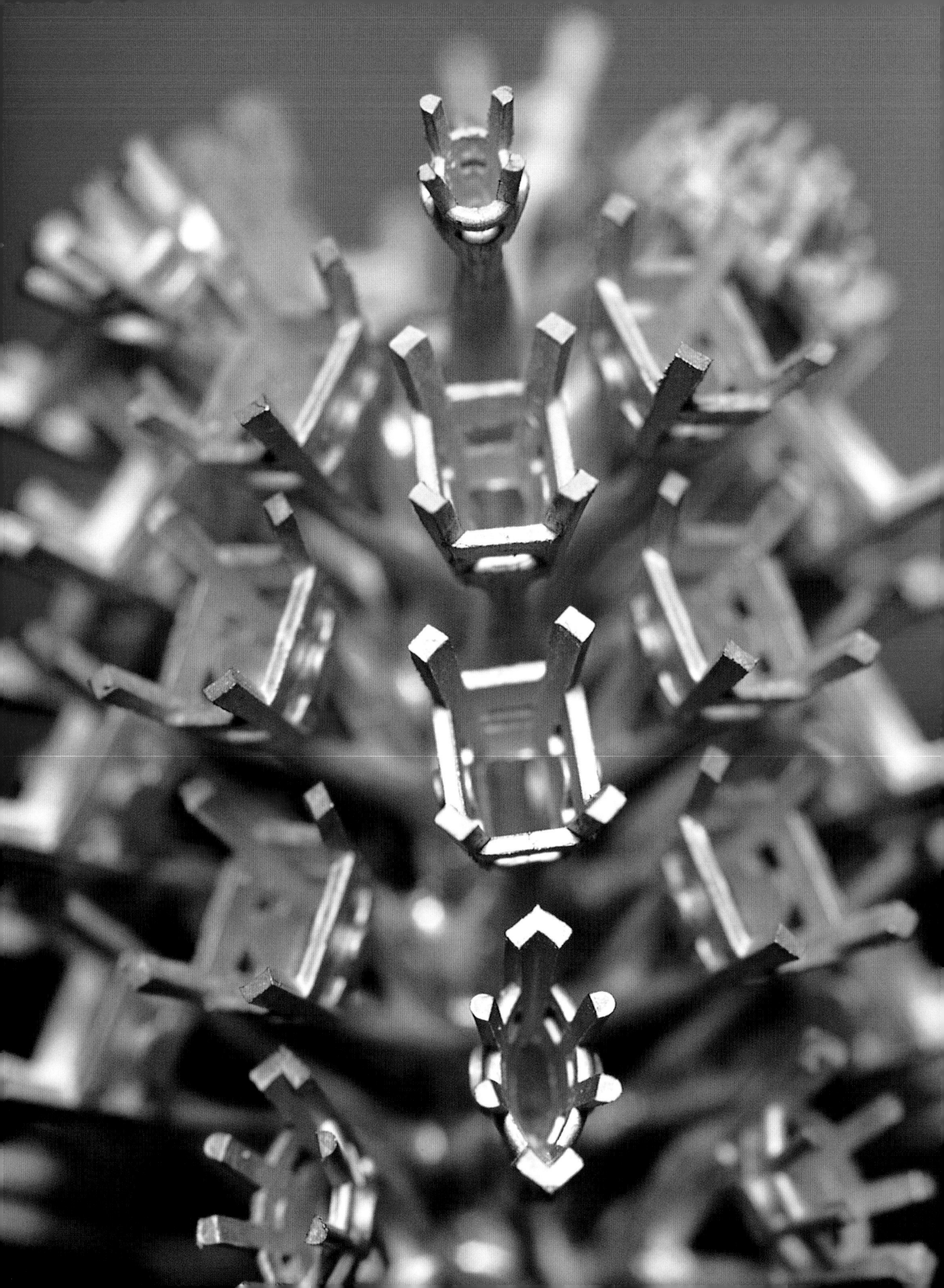

# Platinum Countdown

## THE TOP 10 STEPS FOR TROUBLE-FREE CASTS

BY CHRISTOPHER J. CART

Do you ever wish you had a few tricks up your sleeve when casting platinum? Speaking from experience, I can assure you that magic won't get you very far—but technique and precision will.

Like all manufacturing processes, platinum casting requires careful evaluations, technical innovations, and systematic process controls. From creating the master pattern to removing investment from the cast pieces, each step of the platinum casting process must be tailored to suit the behavior of this tricky white metal. Because if you treat platinum as if it were gold or any other alloy, you're bound to end up with some rather unfavorable surprises.

By adopting the following 10 steps into your casting process, you can avoid these unwanted surprises and instead be on your way to successful platinum casting. Now, on with the countdown…

## 10 Keep your eyes on the sprue.

If the master pattern has been designed for platinum, it can be made substantially lighter than a master made for gold or silver, without compromising structural integrity. Platinum alloys are dense and have specific gravities between 20 and 21.5, as compared with the significantly lower specific gravities of sterling silver, 14k gold, and 18k gold (all of which range between approximately 10.4 and 15.5). Thus, platinum alloys have a structural integrity superior to that of gold or silver, especially at thin cross sections such as prongs or gallery work.

However, special care should be exercised to avoid patterns that go from thick to thin to thick again. Such patterns can cause a Venturi effect in the molten platinum: As it enters the thin section it compresses, and as it exits, the molten metal sprays into the thick section. This traps air in the solidifying platinum, leaving the castings rife with gas porosity.

Additional problems occur as the dissimilar cross sections start to solidify. The thin section cools first, blocking metal flow and creating shrinkage porosity in the thicker sections. Therefore, if the design calls for thick-to-thin-to-thick sections, make a multiple-part master to compensate for such limitations.

As in all types of casting, the master pattern should be reviewed for the best possible sprue placement that will not inhibit mold making, mold cutting, wax injection, or metal flow. For example, if a bezel head is sprued perpendicular to the bezel wall, mold pressure during wax injection can distort the piece; on a square bezel, the wall closest to the sprue will be thinner than the other walls. Instead, the bezel should be sprued at the base of the wall, so that the pressure is directed up into the piece and not against the wall. When casting with hand-carved or handmade waxes, take special care: Since these waxes are often one-offs that haven't been cast before, you can rely only on past experience with similar products to decide where to place the sprue or sprues and what flask temperatures to use.

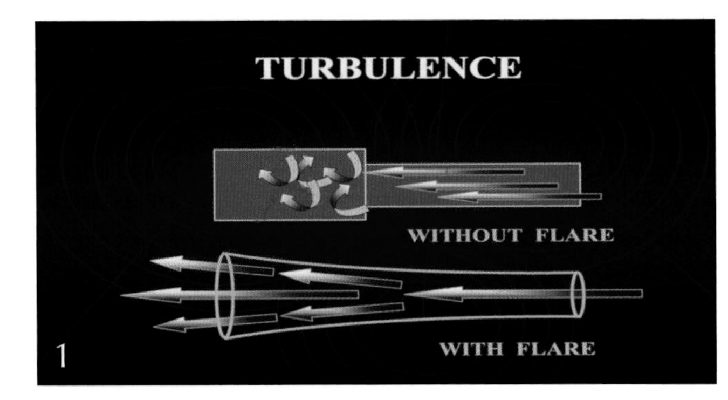

1

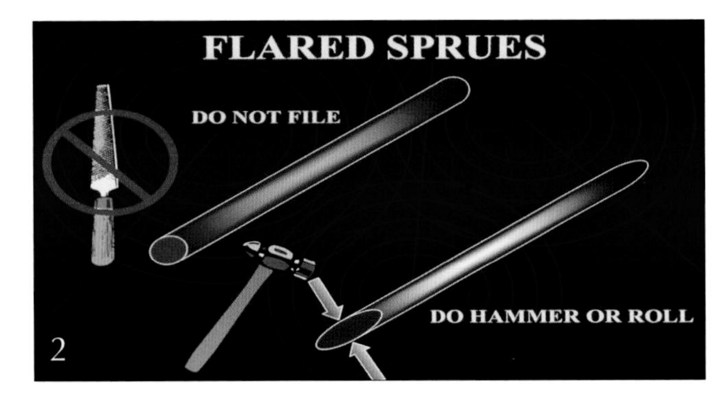

2

In addition to positioning the sprues properly, you should also evaluate the sprues themselves. A sprue that may suffice for 18k gold may not be adequate in all cases, since platinum generally solidifies more quickly than gold. Consequently, additional feed gates may be necessary to ensure the best flow of molten metal into the investment cavity. Always check the melting and casting ranges of the specific alloys used before attaching a sprue.

As a further aid to metal flow, sprues should gently taper toward the pattern whenever possible, allowing for a smoother introduction of molten metal into the investment mold. This will reduce turbulence, which can cause sharp corners in the investment to break off, contaminating the metal and encapsulating air (Figure 1). However, you should not file the taper, since filing will change the volume of the sprue. Instead, hammer or roll the end of the sprue prior to joining it to the master pattern (Figure 2).

And do not forget the most common problem faced by a platinum caster: a hole under the sprue. This hole is caused by shrinkage: When platinum contracts during cooling, the diameter of the sprue must allow in enough molten platinum to fill the void. If it doesn't, a hole forms under the sprue (the last area to solidify). To fill a platinum casting, the volume of the sprue should be at least 1 to 1.5 times that of the thickest cross section to be cast.

## 9 Have seemly, seamless molds.

To ensure perfect representation of the master pattern, you must sufficiently pack the mold frame with un-vulcanized rubber. You should have about 6.5 ply of rubber for every $^{13}/_{16}$ inch.

When cutting the rubber mold, cut along the lines of the master pattern so as to hide or disguise any seams. For example, if you are cutting a ring, you don't want the seam on the center of the shank. By cutting along the side of the shank, you can hide the mold seam on the edge.

You may also need to cut vents to provide adequate airflow from the mold cavity during wax injection. To further enhance airflow, apply talcum powder sparingly to the air vents using a fine artist's brush. Take note, though: Talcum powder should not be used as a parting agent in the rubber mold for the injected waxes. If talcum is present in the injected wax, you can expect a rough as-cast surface, as well as possible contamination of the metal due to residual talcum powder in the burned-out flask. Instead, spray silicone mold release directly onto the mold cavity.

Prior to production, the mold should be tested to establish parameters for a successful injection (air pressure, dwell time, and vacuum) according to the chosen wax. These parameters should then be listed on the mold, enabling an operator to make the proper adjustments to the wax injector prior to injection.

*One additional note:* Hand-carved or handmade waxes should be processed into an RTV silicone rubber "cold mold," which does not require vulcanizing to set. Because the liquid silicone rubber flows and sets around the wax pattern, the original carving can be easily cut out whole. This allows for duplication in case of a poor cast.

## 8 Make a wax and check it twice.

The wax you choose to inject into the mold is usually dictated by the type of piece you are making. For example, if you are casting a piece with fine filigree, thin-pronged heads, or thin cross sections, you should use a wax with a higher fluidity, such as pink wax. On the other hand, if you are casting a large piece with thick cross sections, you should choose an injection wax that has a low shrinkage factor, such as dark green. If you are unsure which type is best for the piece, use a general-purpose injection wax—most are in the blue range. Just remember to adjust the injection parameters for each mold.

Once you've chosen your wax, ensuring a quality pattern can be summarized in one word: inspection. As most casters know, removing mold seams from an as-cast piece is not an efficient use of time or precious metals. And if the wax pattern contains air bubbles, the air can be drawn out during investing, causing subsequent damage to the wax and creating a poor mold cavity and casting.

To prevent such problems, waxes should be inspected twice, first by the wax injection operator and again during the spruing process. The inspectors should make sure that the pattern is free of distortion, mold seams, and air bubbles. Design-specific inspections should also be conducted; for instance, on a wax for a prong-set ring, you should check that the prongs are intact.

Sometimes, it may be necessary to repair waxes. Try using wax polishing fluids, which can smooth surfaces that display evidence of mold seams, and disclosing wax, which can fill voids caused by air bubbles.

## 7 Branch out when spruing your tree.

To properly place wax patterns on a sprue tree, begin by segregating them into several categories: very-fine to fine, fine to medium, medium to heavy, and heavy. With ladies' rings, for example, those weighing from 3 to 5 grams would be in the fine to medium range, and those weighing 10 to 15 grams would be in the medium to heavy category.

Place the very-fine to fine pieces at the top of the tree, the fine to medium pieces in the middle, and medium to heavy pieces at the base (Figure 3). This not only creates stability, but also reduces the risk of investment cracking due to the force needed to fill larger pieces at the top of the tree, a problem that can lead to finning and possible loss of metal after the investment is washed off.

You can obtain even better results by placing similar products on one tree and designating an oven temperature for the casting flask. In a best-case scenario, you should weigh the waxes and segregate the product according to average weights and geometry. This promotes thermal equilibrium during burnout, casting, and the cooling process prior to investment washout, since the degrees of thickness and the weight of the pieces greatly affect their ability to hold heat.

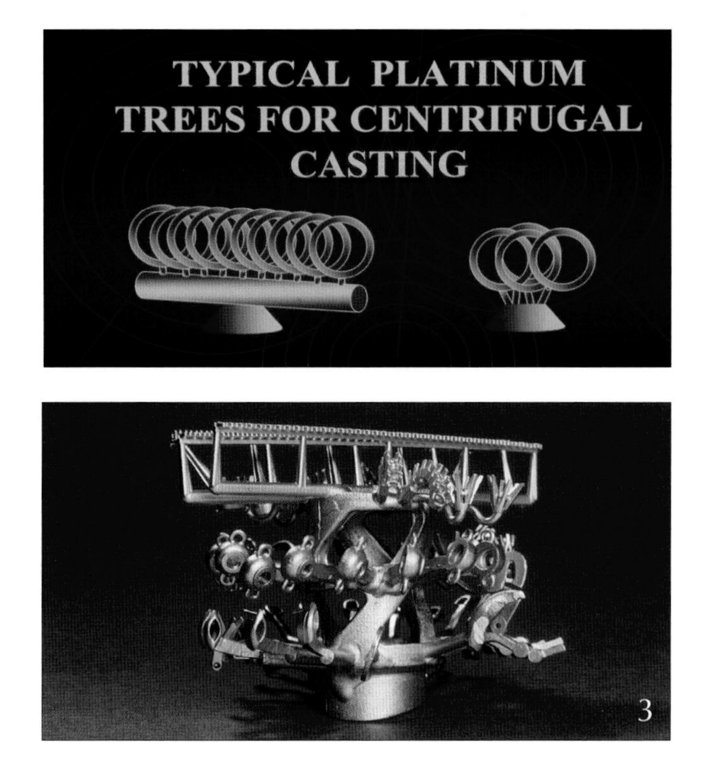

It is also necessary to make sure that you achieve a high ratio of product weight to sprue (scrap). You should always try to apply at least 60 percent product weight per flask. However, don't crowd product on the tree, since this can lead to other problems: Investment can break down, and pieces may touch each other and become welded. The pieces can also burn or develop rough surfaces due to the thermal mass of the products not being able to dissipate the heat properly.

## 6 Invest wisely.

There are several types of investment used for platinum casting, all of which comprise high-temperature refractory silica sand with proprietary constituents. The three most common types of investment for platinum casting, as well as some usage guidelines, are listed below.

**Phosphate Bonded.** Due to its relatively quick setup time (two hours), phosphate-bonded investment is commonly used for casting platinum. A chemically set material, it can be poured into a flask with a rubber base, as in gold casting. Because phosphate-bonded investments have a high viscosity, you should use a heavy-duty mixer when combining the powder and water to produce an investment slurry.

**Phosphoric Acid Bonded.** When used correctly, phosphoric-acid-bonded investment produces superior results. It is a strong sintered investment that provides a very smooth as-cast surface on platinum. However, since it is a sedimentary material and not chemically set, this investment requires 12 to 15 hours to properly set up.

The investing process begins with positioning the flask on an absorbent material (cardboard, heavy-duty paper towels, disposable diapers, etc.) and holding it in place with molten or soft pliable wax. (An absorbent liner is also usually placed inside the flask.) Because this material is sedimentary, it can be mixed indefinitely; the longer it is mixed, the better. However, it will slump in the flask during settling. To offset this, place masking tape around the top of the flask to allow for approximately 0.5 inch of over-pour (Figure 4).

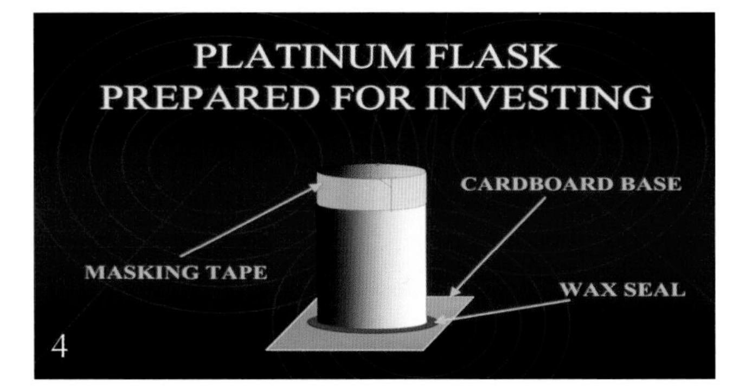


During the setup of phosphoric-acid-bonded investment, you should induce evaporation by creating heat and air circulation. Place the flask in a temperature- and humidity-controlled room set at 74°F to 78°F (23°C to 25°C), with 36 percent humidity or lower. As another option, try placing the flask on the base of a seed germinator (available in seed stores or plant nurseries) set at maximum temperature. To promote airflow, do not put the top on the germinator.

**High-Speed Dental Investment.** A recent adaptation from the dental industry, this investment can go from mixing to burnout to casting in about three hours. This is especially useful when urgent platinum castings are required, since you can invest, cast, and finish a product in one day.

This high-speed investment requires no metal flask; instead, it comes with a silicone rubber base and a tapered, flexible flask that is easily removed. During the set time, a chemical reaction occurs that raises the temperature of the investment to as high as 120°F (49°C), which subsequently starts melting the wax and eliminating residual moisture prior to burnout.

Regardless of which type of investment you choose, following a few general guidelines will put you on the road to a successful cast:

• Control and record all parameters of investment mixing. Document the following: investment type and batch number, investment temperature, water-to-powder ratio, water temperature and pH, slurry temperature and pH, and working and set times.

• Use distilled or de-ionized water when mixing investment. Because of its purity and neutral pH, it will not adulterate the chemical balance of the investment powder or liquid binder.

• Always adhere to the investment vendor's specifications when mixing investment. It's a good idea to mix a sample of each new lot of investment prior to putting it into production.

## 5 Stay cool over burnout.

The process of burnout is normally achieved overnight with the help of a temperature-ramping control device, which is connected to a thermocouple placed in the furnace. Since this means burnout often occurs when you are not in the shop, you should connect the temperature control to a chart recorder; the collected data can be used to correct problems such as investment breakdown caused by an over-aggressive temperature increase or insufficient burnout.

A more accurate form of burnout control and documentation is the use of programmable logic controls (PLCs). PLCs process electrical energy by percentage values, giving a much smoother temperature ramp than standard ramping devices. (Standard controls give 100 percent energy to the furnace, causing an over- or under-ramp in temperature during the burnout process.)

PLCs also allow for data capture that can be downloaded to a computer for graphs and analysis, creating a historical database for evaluating any anomalies that may occur in burnout over time. In addition, some PLCs provide automatic restart in the event of power failure, starting the burnout program at the last stage prior to the interruption.

In addition to recording data, following the proper procedures for loading the furnace is essential. The flask should be placed in the furnace with adequate space to allow for good air circulation and a balanced thermal equilibrium. An overcrowded furnace can cause inadequate burnout due to excessive smoke and insufficient oxygen. (Oxygen is required for proper burnout because it burns off the residual ash from the wax.)

*Note:* There is quite a bit of controversy over whether to use a gas or an electric burnout furnace. Although both can provide adequate results, I recommend using a gas furnace, since it provides enhanced oxygen airflow.

## 4 Keep it clean.

As with all casting, use a high ratio of fresh metal to scrap metal (70 percent fresh to 30 percent scrap). Make sure that all scrap metal is free of foreign particles, such as investment, oxides, or acids. Not using clean metals will inevitably lead to excessive casting failures, which can manifest as gas porosity, brittle castings, and discoloration in the finished jewelry item. If a metal is suspected of contamination, it should be purged immediately from the casting cycle. All crucibles used for the suspect metal should also be removed to isolate the contamination.

## 3 Keep up on meltdowns.

A proper melting procedure includes crucible inspection, temperature measurement, and documentation. Prior to using them, you should inspect crucibles for structural problems (insufficient wall thicknesses, eroded pouring nozzles, cracks, etc.) as well as any residual metal and foreign materials that could cause contamination. You should also document the number of melts produced by each crucible, and create a database to identify and establish at which point the crucibles have reached ultimate degradation.

It's also important to accurately measure melting temperatures, using either a single-spectrum or a dual-optical pyrometer. These temperature-control devices should be upgraded and calibrated on a bi-annual basis. Be aware, though, that optical pyrometers may not work well with platinum/cobalt or platinum/copper. These alloys tend to produce greater smoke emissions and dross across the top of the molten metal, which can affect the temperature reading.

## 2 Cast away.

There are three types of casting machines commonly used for platinum casting. Just as the capacity and features of each machine vary, so too do their requirements for a safe and successful platinum casting.

• **Horizontal or Vertical Centrifugal/Torch** (accommodates up to about 3 oz./85 grams of platinum). This is the most common method of platinum casting for small workshops, although it is the least accurate for quality and process control. Before casting platinum using a torch and centrifugal casting machine, you must balance the arm of the centrifuge. First, place the flask to be cast in the flask saddle on the casting arm. Next, adjust the weight on the casting arm until it becomes horizontal and parallel to the casting machine base. Failure to balance the centrifuge arm causes it to undulate, which can result in metal spilling or spraying around the workshop. Worse, the flask can become dislodged from the rotating arm and be launched across the shop, possibly causing grievous bodily harm to anyone in the room.

To avoid possible contamination, it is recommended that hydrogen gas with oxygen be used for torch-melting platinum. However, either oxygen and natural gas or oxygen and propane gas can be used if sufficient air pressure is applied at the oxygen pressure regulator. When the metal is molten, you must be very quick in releasing the arm and pulling the torch away (up for horizontal centrifuge and out for vertical centrifuge). For this step in the process, the vertical centrifuge offers advantages. First, an increased g-force is produced when the centrifuge arm and crucible rise in the first quarter turn. Second, if there is a spill or a flask becomes dislodged during casting, the hot metal or flask is thrown up and away from the operator.

• **Centrifugal/Induction** (accommodates up to 14 oz./400 grams of platinum). This method, which uses an induction coil to melt the metal, offers a more advanced melting technology than the centrifugal/torch method. Induction heating or melting occurs when electrically conductive materials, such as platinum, are immersed in an alternating magnetic field. (An electrical coil that is energized by a suitable AC generator produces this field.)

This is a cleaner and quicker method of melting platinum than torch melting, with less dissipation of heat.

A properly engineered centrifugal/induction machine incorporates a dual broken-arm assembly. This assembly compensates for the weight of the platinum (inertia) and centers the molten metal into the waiting spinning flask (Figure 5). However, as with the centrifugal/torch method, this system likewise employs a centrifuge; turbulence is an issue, and you must balance the flask.

Fortunately, the newer centrifugal/induction casting machines incorporate vacuum assistance, which removes air from the chamber and the mold itself. This helps to reduce turbulence by eliminating the resistance of the molten metal to any air that may be trapped in the awaiting flask.

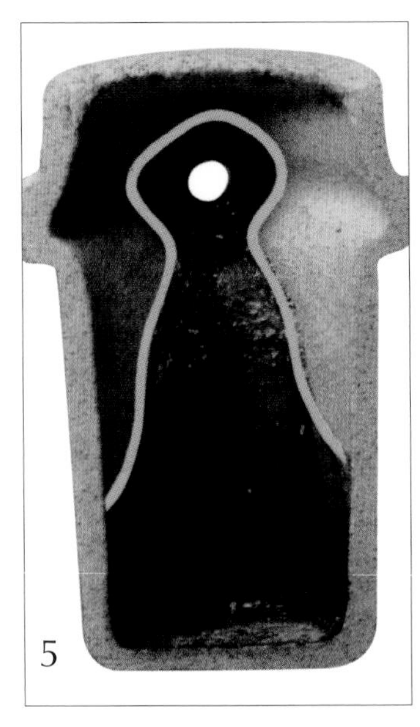
5

However, no matter how well made, most centrifugal casting machines basically "sling" metal into the mold, and thus offer the operator limited control of the casting operation. Also, because so much force is required to adequately fill the mold cavity, the casting tree cannot weigh more than 400 grams or so, thus reducing output. (The usual output for this type of system is a tree weighing about 150 to 250 grams.) To overcome these inherent limitations, casters must look toward the "next generation" of casting machines.

• **Next Generation** (accommodates up to about 25 oz./700 grams of platinum). These high-temperature induction-melt casting machines exploit both gravitational and centrifugal forces: They incorporate tilt pour, and the vertical flask rotates from 0 to 600 rpm, promoting continuous gas substitution to molten metal. They also feature both vacuum and overpressure (through the introduction of argon gas), and supply alternating push/pull forces, as opposed to the singular force of the centrifuge. Add in a dual-spectrum optical pyrometer for accurate temperature readings, as well as programmable logic controls, and you can see how these newer machines take many of the variables out of platinum casting.

Not only do these machines promote better control and efficiency, they also increase productivity: Whereas standard centrifugal casting machines accommodate 125 to 400 grams of platinum, some of the "next generation" platinum-casting machines have a crucible capacity of up to 700 grams. For many casters, this increase more than makes up for the higher price tag.

## 1 Blast off residual investment.

Regardless of which investment you choose for casting platinum, all are more difficult to remove than the gypsum-bonded investments used for gold casting. Because the high flask temperatures required to cast platinum (1,000°F/538°C to 2,000°F/1,093°C) and the melting temperatures of platinum alloys (which can often reach up to about 3,236°F/1,780°C) cause the investment to fuse to the platinum casting, special methods are necessary to remove it.

After casting, immediately blast the hot flasks with high-pressure water to free the cast platinum tree. To remove additional residual investment, use high-pressure wet bead blasting and submerge the trees in hydrofluoric acid overnight; this will remove any remaining oxides or investment. Other baths can be used as a safe substitute for hydrofluoric acid, such as a mixture of 25 percent sodium hydroxide, 25 percent potassium hydroxide, and 50 percent water, or an ammonium bifluoride and water mixture comprising 8 oz. of ammonium bifluoride to 1 gallon of water.

After the bath, neutralize the cast trees in a bicarbonate of soda solution. Next, submerge them in an ultrasonic bath of water and ammonia to neutralize any residual acid. Remember: All devesting should take place under OSHA-approved air extraction conditions. Devesting operators should wear protective clothing and eyewear, especially when entering the hydrofluoric acid bath area. Hydrofluoric acid is an extremely dangerous material that requires a specific emergency kit in the event of exposure.

If the devesting has been done correctly—and all of the other nine steps have been followed—you should have a tree full of clean, smooth platinum castings. Now you can once again begin counting—your profits, that is. And that's no trick.

# Beating the Bi-Metal Blues

## HOW TO BOND PLATINUM AND GOLD SEAMLESSLY

BY JURGEN J. MAERZ

"What a beautiful design," you say, admiring the rendering of a modern, two-tone ring. The top of the shank splits into four tapering, complementary prongs—two yellow gold, two platinum—that will hold a single stone. The bottom of the shank flares and is divided evenly between the two metals, which taper and mesh like a pair of folded arms. The designer has obviously put a lot of effort into this piece; now it's up to you to figure out the best way to make it.

You sit back in your chair and study the design further. You've always liked the combination of platinum and 18k—the great strength and shine of platinum, coupled with the warmth of yellow gold. What a combination! And, if you play your cards right, the design will sell in fairly decent quantities at the stores you've been dealing with for years.

So you create the model. It's exciting to see the pieces fit together, the two parts that will join to make this great ring. They fit perfectly, and once again you admire the design: the sleek lines of the shank, the way the prongs rise to hold a stone... You finish the model and are pleased. You select a spot for the sprue on each of the two halves and vulcanize the pieces in a rubber mold.

After cutting the mold and injecting the wax, you cast three of the platinum pieces and three of the gold. Finishing the platinum and reducing the scratches take a little more time, since the metal's hardness and toughness require special polishing compounds. But you realize that the extra work will help you later, since gold polishes faster and can be easily over-polished once it is joined to the platinum.

Next you solder the platinum and gold sections together. Satisfied with your work, you take a closer look at the seam—and your heart skips a beat! Your eyes won't believe it, but there they are: fine cracks in the platinum.

Oh no! you scream inside. Taking a closer look, you find more cracks along the entire seam. Stress! You must relieve the stress. Platinum expands at a different rate than gold, you realize, and because the solder is already in place, the piece is cracking and warping.

Heat should help. You solder two more parts together and place the piece in a preheated kiln. After 30 minutes at 1,292°F (700°C) it should be perfect. You turn off the kiln and let the ring cool inside. As you take it out, you look at it and smile. It looks good! You finish the piece—and there they are again! Cracks! The trouble is, they don't show up until you are almost done with the finishing. There is no way to burnish, no good way to fill.

You want to tear your hair out. You sit back, look at the design, and think: What to do? What to do?

Suddenly you jump out of your chair: Why not cast the gold onto the platinum! That way you can avoid soldering. Besides, nothing is going to hurt the platinum.

You take the third platinum part, lapping and completely polishing it. You then inject a wax mold for the half of the ring that is going to be gold, and carefully attach the wax model to the platinum ring. You also make sure the wax is a tiny bit larger than the platinum all the way around, to compensate for the polishing that is yet to be done on the gold, as well as for potential shrinkage.

You find the place on this assembly where it makes sense to place the sprue, and invest the entire package. Under the vacuum bell the bubbles rise to the surface. Is this going to work? you wonder. You de-wax and place the flask into a kiln.

All night you let the burnout cycle do its thing. The next morning you race to work and find the flask glowing a bright red in the kiln. You take the 18k grain, melt it, and cast. Inside the flask the liquid gold races toward the hot platinum. It splashes against the polished surface, and at the grain level a metallurgical bond takes place. The gold literally cuddles the platinum in the flask.

You remove the investment and look at the product of your imagination. It looks great! You jet-spray the casting and loupe it. The ring is perfect: no cracks, no warping, just a bi-metal casting with a brightly polished platinum side. This technique is going to open up an entire new world of possibilities, you think. You finish the gold portion with a smile, set the stone, and look at the ring.

"What a beautiful design," you say...

# A Big Idea

## CASTING 570 GRAMS OF PLATINUM

BY JURGEN J. MAERZ

One of the perks of my job as technical director for Platinum Guild International (PGI) USA is the opportunity to learn new methods for working with platinum. It's exciting to be involved in developing breakthrough techniques that make a difference in the industry. But even more exciting than learning this information is the opportunity to pass it on to other manufacturers and watch the processes evolve.

One such opportunity materialized the day Michael Epstein of EPS LLC in Feasterville, Pennsylvania, came to the PGI booth at the JA New York show. He wanted to discuss a new platinum casting technique that he claimed yielded up to 65 rings in one casting. I must admit, at first I was a bit skeptical. Everything that I had ever learned about platinum casting told me that this simply could not be done. But seeing is believing. When he reached into a satchel and slowly unwrapped the biggest platinum cast tree I have ever seen—I believed.

Measuring more than 6 inches tall, it had about 65 eternity bands on a very thin center stem. The rings were tightly stacked and connected by an amazingly thin sprue (about 1 mm long by 2 mm wide) to a center stem that was only 4.5 mm in diameter. "This can't be," I thought. The average casting tree is no more than 4 inches tall, contains 10 to 20 rings at most, and generally yields under 250 grams of platinum. Here I was seeing over 350 grams (225 dwt.) of platinum on a type of tree one usually sees in gold casting.

From our initial conversations, I learned that Michael is originally from Russia and had been casting platinum for only about one year. He applied methods he knew from gold casting and just did them. Because of his lack of experience, he did not know what traditionally hasn't worked; he just tried it his way and tweaked it until it came out satisfactory. And what a job he did.

His rings contained almost no porosity, and the trees had a fill of over 95 percent. Sometimes, he said, there were minor no-fills on the top of the tree, on the bottom, or in the middle. But even if you take these no-fills into account, the net yield outperformed anything I've ever seen (with the possible exception of some of the new pressure-over-vacuum machines, which have a higher capacity for melting in much larger crucibles).

Considering the rate at which platinum freezes and the time it would take to reach the top of the tree, such a large casting would seem impossible. But clearly, it wasn't. And I was curious to see how it was done.

Michael agreed to show me his procedure for a single casting of 65 platinum rings, and I extended my stay in New York City to visit his factory. It is small, clean, and well organized. From the machine shop where he makes tools and dies to the assembly station for bench jewelers to the casting/waxing room, the work areas are well designed and functional. He has a large Galloni Modular 6 induction casting machine and an electric kiln for burnout.

In anticipation of my arrival, Michael had prepared four flasks for casting. (He uses Platinum Plus, a two-part phosphoric-acid-bonded investment manufactured by Ransom & Randolph in Maumee, Ohio.) He then proceeded to show me the entire process, explaining each step.

## Investing

Platinum Plus is prepared like many two-part phosphoric-acid-bonded investments: The binder concentrate is mixed with water to create the solution, which is then mixed with the powder to make the slurry. As the setup dries, the water must be removed to allow the acid to interact with the investment powder, making an exact replication possible. This is commonly achieved by using a paper liner inside the flask, and then setting the flask on a cardboard base, which aids in water removal. The investment turns into a gel-type substance that settles during the drying process.

In his initial experiments with Platinum Plus, Michael noticed that cracks often appeared as the investment settled in the early stages of the drying process. After much trial and error, he found that by diluting the manufacturer's recommended ratio of acid and distilled water in the binder by 30 percent, he could create a more porous and less dense investment setting. (The new ratio was 1 part binder concentrate to 26 parts water.)

This has many advantages. During casting, as the liquid platinum enters the cavity, the gases can escape through the less dense pores of the investment, making it possible for the metal to travel rapidly and fill the mold. The investment is not as hard, and devesting is less complicated.

In addition, Michael uses a perforated stainless-steel flask that measures 3 inches in diameter by 8 inches high by 0.125 inch thick. The use of a perforated flask in casting is not new; however, it is rarely used for centrifugal casting. On a tall flask, it is difficult to remove the water through the bottom. By using a perforated flask and wrapping the outside with several layers of paper towels, the water can be removed slowly and evenly.

In trial castings it was determined that the best results were achieved with a 4.5 mm center sprue for a single tree, and a 3 mm center sprue for a triple tree. To create a wax stem that is thin yet still strong enough to support 65 eternity rings, Michael dipped a 1.5 mm copper wire into the wax pot and coated it to make a 4.5 mm sprue. (This takes a bit of practice.) The rings could then be waxed on the tree, and the tree waxed to a domed base. (After the wax melts away during burnout, the copper wire can be removed easily.)

## Burnout

After pouring the investment into the wrapped flask and vacuuming to remove air bubbles, the flask remained on the bench for one hour. It was then set in the kiln, which was ramped to 200°F (93°C) in 30 minutes, and held at that temperature for two hours. Next, the temperature was brought to 350°F (177°C) in 30 minutes and held at that temperature for one hour. At that point the kiln was opened, the paper and the copper wires were removed from the flask, and the flask was placed back into the kiln. The temperature was then ramped to casting temperature—between 1,400°F to 1,700°F (760°C to 927°C).

## Casting

According to Michael, the first step in achieving his amazing castings is to reduce the speed of the casting machine to between 200 and 300 rpm. Many casters assume that higher speeds yield better results when casting platinum. This is not true. At high speeds, the metal enters the flask with turbulence. Any sudden change in direction or gas obstruction will shorten the distance the liquid platinum can flow. By reducing the speed of the machine, the metal enters in a smooth, even flow; it pushes away gas inside the flask and fills the cavity with clean, porous-free castings.

Note: For large pieces, such as bracelets, Michael recommends a casting arm speed of 200 rpm. For most rings, Michael obtained the best results with 250 rpm. For a triple tree casting with many small parts, 300 rpm worked best.

To begin casting, Michael placed a 1,700°F (927°C) flask in the cradle of the machine and filled the crucible with a small amount of platinum casting grain. The induction coil was activated. As the platinum started to melt, more grains were added to the melt until all 360 grams were in the crucible (the maximum capacity) and liquefied.

The metal temperature was 3,542°F (1,950°C), which represents a super heating of about 392°F (200°C) above the liquidus of the platinum alloy used, platinum 950/iridium 50. I personally do not care for this alloy for casting, as it is rather soft (about 80 HV). However, since Michael was casting eternity rings that were to be set with stones all around the band, a softer metal was preferable. Michael claims that with his process, the metal seems harder than other platinum 950/iridium 50 alloys that he has worked with. (This is an opinion that I cannot confirm. However, I believe that the hardness of these rings is increased during burnish setting.)

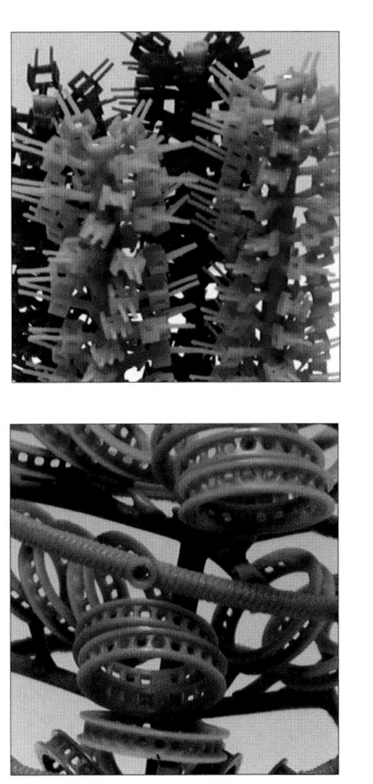

When the flask was removed from the cradle, it looked like a good cast. The button had a nice dip in the center, and all the metal was inside the flask. There were no spills and no cracks in the investment.

Michael set the flask on a fire brick to cool. Then, while intermittently dousing the flask in a bucket of water, he used a hammer to remove the casting. The tree was amazing. He let it soak in a bath of hot 45 percent caustic potash for about 30 minutes, then removed it from the stainless steel pot. All the investment was totally gone, the platinum was shiny, and the castings were good. The rate of no-fills was about 5 percent.

## Experimentation

In the shop that day, Michael performed three additional casting experiments. The second tree with 65 eternity rings had a hollow square center sprue; no copper wire or wax was used as in the first experiment. Amazingly, the tree filled all the way to the top. There were a few more incomplete castings than before—about 15 to 20 percent compared to 5 percent on the previous tree—but because of the high yield, the casting was still successful. With a little more experimenting, I am convinced that the castings with the hollow centers can be improved.

The third tree had three center sprues, each of which measured 3.5 mm in diameter and over 6 inches long, coming up from the cone. (These center sprues were made with copper wire for durability.) The parts for three complete line bracelets, comprising 55 links and one lock each, as well as some parts for a ring were waxed onto the three main sprues. With the casting machine speed set at 300 rpm, the result was a complete fill.

His fourth tree, which contained three complete one-piece bangle bracelets, also produced good castings.

Seeing no reason to waste metal, Michael used the feeder sprues to the bracelets as sprues for additional small parts. At 350 grams (225 dwt.), the tree yielded 270 grams (174 dwt.) of usable product. In a later casting, after some modifications to the machine, a 550-gram (354-dwt.) tree yielded 402 grams (258 dwt.) of usable product.

Like most experiments, Michael's platinum casting technique does have its problems. But I believe they can be solved with research. Take, for example, the problem of no-fills. This is due in part to the Galloni induction casting machine that Michael uses, which has a double broken-arm system. During initial acceleration, the platinum favors the outside of the tree in the opposite direction of the acceleration. This leads to an interruption of the flow, which can make the platinum freeze and thus not fill. Working from a theory that a smooth acceleration and a change in the placement of the tree within the flask could compensate for this problem, Michael invested a few trees and bent the center sprue about 10 to 15 degrees to the side. The results were phenomenal.

Furthermore, since our first meeting, Michael and I have done extensive casting experiments with various platinum investments. We tried casting with Opticast from KerrLab in Orange, California; J Formula from Romanoff International Jewelry Supply Corp. in Amityville, New York; and Platinum QC from Westcast in Albuquerque, New Mexico. With each of these investments, reducing the acid content by varying degrees produced successful castings.

In an attempt to cast more than 500 grams (322 dwt.) of platinum product, Michael exchanged the coil of the Galloni with the company's larger-size coil and added a larger crucible. This made it possible to cast over 500 grams (322 dwt.). The cast trees weighed between 350 grams (225 dwt.) and 570 grams (367 dwt.).

To prove that this casting method is suitable for heavier items, we cast a tree of platinum frogs that weighed 1 oz. each. We also successfully cast a tree with four main sprues containing all the parts needed to make four complete line bracelets. This tree resulted in about 3 percent no-fills.

But discoveries such as these are only the beginning. Michael has agreed to try his methods of casting with a variety of different options. In addition, I encourage other casters to test Michael's methods in their own shops. Because while it may be my job to pass exciting new techniques on to you, it's your job to continue experimenting with them—and sharing the results of your research with the industry.

# Curing Altitude Sickness

## YOU CAN'T FOOL MOTHER NATURE...CAN YOU?

BY MARC "DOC" ROBINSON

Almost 20 years ago, I was doing some consulting work for a large casting operation in Albuquerque, New Mexico. When I blew into their scene, they had a huge table of workers armed with needle-nose pliers, diligently pulling nodules off of trees that had been caused by air bubbles in the investment. This was an actual process within the scheme of their casting theater, created to eliminate a problem that is normally dealt with during the investing procedure by vacuuming.

Why so many bubbles? And why did the vacuum process fail to eliminate them?

The answer hinged on that company's location. All casters in high altitudes have problems with vacuum machines—the higher above sea level you are, the fewer inches of vacuum you will be able to draw. In Albuquerque, which is nearly a mile high, you could draw only 24 to 25 inches maximum. This is in comparison to a full 29.9 inches at sea level. As a result, this company could not draw enough vacuum to effectively deal with all the bubbles clinging to the waxes inside the flasks.

This is strictly a matter of physics, and everybody knows you cannot fool with Mother Nature...or can you?

I reasoned that if there were a way to lower the vacuum boiling point so that the mix would boil fully a lot sooner, we could lick the problem of vacuum vs. altitude. And the way to do this was to raise the temperature of the water in the mix.

So their mix had to be readjusted. They had to raise the water temperature, lengthen the mixing time, and generally thin out the investment...relying on the fact that when you up the water temperature, the investment sets up very, very fast. Much too fast to rely on any standard room-temperature formula.

In my book, the faster the investment glazes over and sets up, the stronger the plaster after burnout, and the less likely you will lose valuable detail from water separation along the wax pieces.

With all this in mind, here's the formula I devised for companies suffering from the dreaded "altitude sickness."

For two perforated 4-inch flasks, 3.25 inches in diameter, I mix 1,020 dwt. of powder to 700 ml of water at 110°F (43°C). This is then mixed at a quick speed for a full six minutes.

The bowl is then placed under vacuum for the first boil. Depending on the strength of the vacuum machine, it can take between 30 seconds and one minute for the mix to initially rise up and boil. I time 15 seconds of boil after the boil commences. I then pour the flasks and vacuum again, this time allowing a full one-minute boil.

Now I pull the bell jar, top off the flasks, and set them aside to time the glaze-over. The setup time should average 1.5 to 3 minutes, total. This is very fast, and ensures detail and strength by preventing the water and plaster from separating.

An advantage to the mix pouring thin is that the investment naturally flows into details and crevasses more completely than a thicker investment, and is capable of mirroring much finer detail. This is coupled with the fast setup time, which, by reducing water separation, likewise helps maintain detail and toughness. Also, when you use hot water, the mix will boil violently—so much so that the bell jar will be totally covered with plaster. This "superboil" not only rids you of bubbles, but also serves to keep mixing the investment during the actual vacuuming process.

The water temperature of 110°F (43°C) at 5,000 feet is admittedly extreme. You may have to go up only a few degrees for your area's altitude. But these principles of hot water mixes even hold true in my California shop, which is near sea level. Of course, if you want to experiment with this process, I recommend practice on an empty flask—investment is cheap compared to wax labor.

That, in effect, is how I fiddled with Mother Nature (sorry, Mom). By raising the temperature of your water, you can induce the investment to boil sooner.

But Mom got me back...after I left Albuquerque, I had to change my name and appearance. It seems there were hundreds of workers in shops there who actually made their living by pulling bubbles off castings! Now they were all after my hide...which brings a new meaning to the term "contract casting."

# Every Stone in Its Place

## CASTING WAX-SET GEMS

BY EDDIE BELL

During his 60-odd years as a master bench jeweler, my father never did a lost-wax casting. Rather, from the time he started as an eight-year-old apprentice in 1908, he did cuttlebone and sand castings to get basic shapes, then invested a great deal of work into making them presentable.

I, on the other hand, wasn't so patient. Although I listened intently as my father taught me how to make and repair jewelry, as a teenager I taught myself the art of lost-wax casting. I say "art" because I didn't realize until much later that casting is best practiced as a technology. But it did occur to me that I could set stones in wax much easier than in metal, and with nothing to inhibit me I successfully cast a jade cabochon in a silver belt buckle.

As time went on, I continued to experiment with casting wax-set stones. The results weren't spectacular, and finishing around the stones sometimes took much longer than usual. But I persevered, experimenting and searching for information.

I was lucky enough to be able to travel to the factories that were also experimenting with this new process—just as it occurred to me to cast on stones, so it occurred to nearly everyone who wanted stones fixed in metal with less effort. Through the collective work of these pioneers, wax-set stone casting gradually became more of a commercial production.

Of course, this transformation is never easy—unexpected problems always arise, and questions must be answered. To help this process along, I'd like to offer answers to some of the most commonly asked questions about stone-in-place casting, in the hope that they will help make the path easier for new pioneers, as stone-in-place technologies continue to break new ground.

### *When casting diamonds in place, what are the most common problems that should concern me?*

There are several common problems associated with wax-set diamonds:

- **Setting stones too closely.** When creating the model, you must remember that, during heating and cooling, metal expands and contracts at a faster rate than the stones. If you set the stones in the wax girdle-to-girdle, as you would set them in metal, they may look nice, but there will be no room for the vise-like contraction of the metal; the diamonds will crush each other. It's best to leave a 0.1 mm space between each stone.
- **Using unsuitable alloys.** Some alloys have been known to contract so much that they cannot be used with stone-in-place casting, regardless of the amount of space left between each stone. The number of alloy formulas prohibit listing definitive guidelines, so it's best to always tell your metal supplier that you intend to perform stone-in-place casting. Your supplier should not only help you to avoid unsuitable alloys, but also point you to possible alloys formulated specifically for stone-in-place operations, such as those with slower solidification rates.
- **Exceeding the recommended burnout temperature.** Flask burnout for diamonds in stone-in-place investment, for example, should not exceed 1,139°F (615°C); exceeding this recommended temperature could cause the diamonds to become cloudy.
- **Quenching too quickly.** Many diamonds are shattered because the flask is quenched while it is still hot: The resulting thermal shock breaks the stones. Always let the flasks air cool for about four to five hours. Before spraying, place the flasks in a bowl of water to quench the button for about 10 minutes.
- **White coatings.** When oxygen reacts with the surface of the diamonds during burnout, a cloudy white coating can appear on the stones. Adding boric acid to the investment can help solve this problem, but it will leave the investment rock hard. For devesting, you will need high-pressure water blasting.

### *How best can I set stones level?*

Setting stones level is a challenge. Even when the wax-set stones are perfectly set, they still might move in the investment.

Success starts with the model. The opening below the stone should be as wide as possible, so that the investment can hold the bottom of the stone after the wax is removed.

In addition, if you don't open up the model, tiny air bubbles can become trapped between the stone and the investment—which means that once the wax is gone, there is nothing left to hold the bottom of the stone in place.

Other causes of floating stones include the use of stones that don't properly fit their seats, investment that hasn't been properly mixed and thus won't set firmly, and wax with too little memory to hold the stones tightly.

### *How should I alter my current spruing and casting techniques for the stone-in-place process?*

Trying to save time by not reworking master models for wax-set stone casting actually wastes time—and money. Feed sprues that are too small, promote excessive cooling of the metal en route to the pattern, or deliver a stream of metal directly on a stone can cause many failures, as they can in all types of casting. With stone-in-place, however, they take on greater importance.

This is because of the lower temperatures required for casting on stone. There is a limit to how hot the investment can be with stones inside: For diamonds, flask temperatures should not exceed 950°F (510°C) when using regular investment, or 1,139°F (615°C) when using stone-in-place investment. The lower burnout temperatures need to be compensated for by doubling the flask soak time at maximum temperature. For most other stones that can be cast in place, normal burnout temperatures can be used with a slower ramp rate.

When higher temperature alloys, such as palladium white gold, are cast into a much lower temperature flask, the metal temperature drops like a rock, and the metal begins to freeze. For this reason, feed sprues must be designed to deliver the metal to the pattern with as little heat loss as possible.

This is especially true when diamonds are cast in place. In industrial casting, chill plates are sometimes installed in the mold to draw heat away and cool a selected area of the casting. Diamonds serve this same purpose, since they are good thermal conductors. The more that the metal comes into contact with a diamond's surface, the faster the heat will be drawn from the metal. Casting such a pattern might require moving the location of the sprue—for example, moving it from the thin base of a ring shank, where metal will tend to freeze first, to a shoulder.

### *I've heard that finishing is difficult with stone-in-place casting. Is this true?*

Since stones can loosen during polishing, the as-cast surface finish must be as perfect and as smooth as possible. Polish all defects off the wax patterns before setting, and set the stones on the same day the wax is injected; a day or two makes a big difference. Magnetic tumblers can help with this task: These machines use media made up of little steel pins that get in behind the stone and make a bright scratch finish.

Finally, you should also look into the type of mold rubber you use: For the most part, vulcanized silicone rubber has replaced natural rubber molds because the quality of the wax pattern is better.

### *Can I perform stone-in-place casting with stones other than diamonds?*

With some you can. Corundum—i.e., ruby and sapphire—is a single-crystal aluminum oxide (chemically, $Al_2O_3$) and one of the most durable stones on which to cast metal; it's hard to make a mistake with it. Unfortunately, not all stones are as friendly—or so it would seem. Some sources say emeralds and opals are not suitable for stone-in-place casting, while CZs are okay. However, I don't remember ever cracking an emerald, although I have cracked a few opals. But lots of CZs have been destroyed. Isn't it odd that two stones considered unsuitable for stone-in-place fared better than one that should be fine?

### *What should I be concerned about when attempting stone-in-place casting with CZs?*

The results from the trees I cast with mixed stone types lead me to believe that CZs are sensitive to thermal shock; cracking can occur when the metal either heats or cools faster than the stone. However, it's rare to crack small, round CZs cast in the same temperature conditions, since these stones usually have less contact with metal than larger stones, and the forces that break the stones have less leverage on the smaller spans.

We later tested these hypotheses on a few beautifully prepared wax-set CZs. After the flask was cast, we put it in a closed container filled with vermiculite to insulate it. After 24 hours, it was still warm to the touch. The result: no cracked CZs. These tests lead me to believe that cracking is more influenced by the cooling cycle than the casting cycle.

### *How can I best determine which stones can be cast and which can't?*

Many times, information about what stones can and can't be cast gets printed and repeated often enough that it becomes accepted, true or not. For example, many jewelers believe that you can't heat an emerald without cracking it. This is based on the fact that thermal gradient cracks are a real danger to many gems, since slow uniform heating and cooling is not practical. However, saying that a non-uniform temperature change in an emerald can cause thermal cracks is much different from saying that you can't ever heat that stone.

I say this based on personal experience. Once, I received two pictures from Hubert Schuster of the Jewelry Technology Institute in Vicenza, Italy, showing a rather large natural emerald that had been wax-set and then cast in white gold (see page 92). Common wisdom would say such a stone-in-place casting can't be done—and here was proof it could be.

However, while doing something one time—or even a few times—may prove it can be done, that doesn't mean we fully understand it. To be honest, I don't know why casting on emerald and opal materials has not cracked every one of them. In the case of diamond, it is such a good conductor that the heat is drawn rapidly away from the source. Also, diamond is not a compound, so all the atoms are the same size in the lattice; perhaps this reduces stress. But with emeralds and opals...I simply don't know why they can be cast successfully.

The only suggestion I can give is to follow all of the basic guidelines listed above, and to recognize that with emerald, opal, or any other gemstone, you can't always listen to the generally accepted wisdom. What stone can and can't be cast? To answer that, all we can do is draw on our experience, examine questions based on science, and experiment—as any good pioneer should.

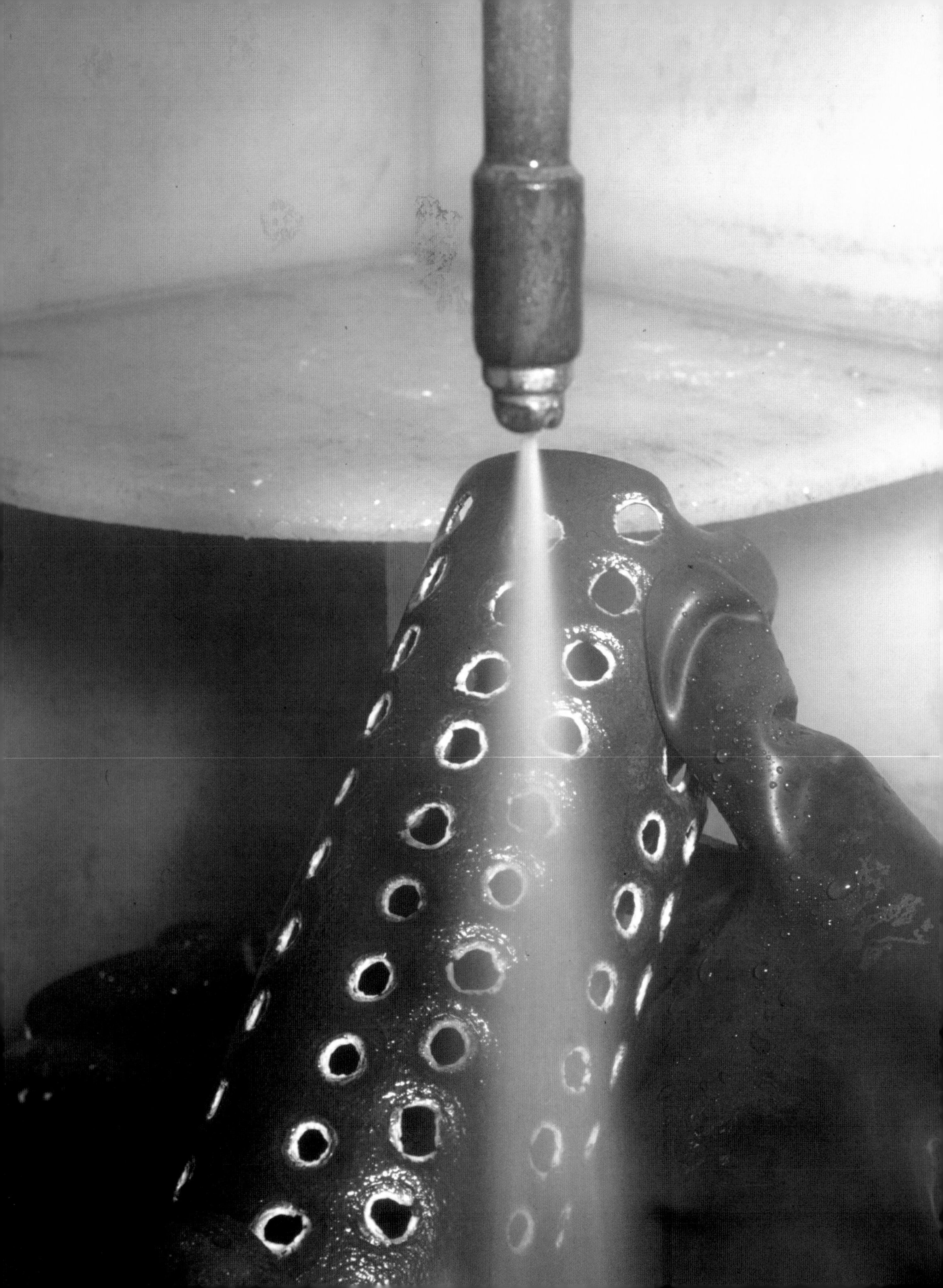

# Breakout Basics

## DETERMINING HOW BEST TO REMOVE CASTINGS

BY GREGG TODD

It's the last act of casting, the step that allows you to see the fruits of your labor. But for many casters, it also raises one last question: What's the best way to break out castings from the investment? Should they be removed hot or after they've been cooled? Should the dry investment be shattered by rapid-fire steel hammers, or subjected to thermal shock by being submerged in water?

These questions would be relatively easy to answer if we used only one metal type and formula throughout the industry. But "easy" is a word that seldom applies to the jewelry profession—and it certainly does not apply to the breakout process, also known as devesting. Instead, in addition to a host of process-specific issues—including flexibility, health risks, and disposal regulations—casters must take into account metallurgical principles.

Like I said, "easy" seldom applies to the jewelry industry.

Let's begin by looking at the varying methods used to shatter investment. Each of them has its advantages and disadvantages, but all can be divided into two basic categories: dry breakout and wet breakout. As you can guess, the main distinguishing difference between the two is water.

## Getting Wet

Many people favor a wet breakout due to its ease and speed. One technique is to repeatedly submerge the hot flask for brief periods in a container of water, until the casting has cooled to near water temperature. A second technique is to simply dunk the flask in water and wait until the temperature equalizes. Both techniques work equally well to aid in the removal of the investment. The first technique results in a little less wet investment, but it can cause hot investment to fly out of the containment area. It also produces more airborne silica if proper ventilation is not provided. For this reason, it should be done only in a properly designed, enclosed area, with the caster protected from flying particles. The second technique cools the casting more quickly, and the investment is contained in the water.

Once the majority of the investment has been broken from the castings, the remainder is removed through a washing operation. It can be as simple as a brush and a bucket of water, or as high-tech as state-of-the-art water-blasting cabinets. The objective here is to remove as much of the residual investment as possible. However, even after this process there may be minute traces of investment left behind. To remove these traces, soak the castings in a pickle solution. Not only does the pickle help to remove the remaining investment, but it also helps to remove oxides from the surface of the castings.

The last steps are to rinse the castings clean of pickle, soak them in an ultrasonic cleaner, and re-rinse to remove any detergents. After that, the finished castings are dried and ready to be removed from the button.

## Drying Out

Dry breakout, on the other hand, relies primarily on vibration to fracture the investment. It can be done when the castings are hot or air-cooled (rather than quenched). Depending on the needs of the particular metal (which will be discussed later), this method gives casters greater flexibility: flasks can be cast and broken out individually, or numerous flasks can be cast in succession and then all broken out at one time.

Dry breakout involves a series of steps, all of which apply vibratory pressures to different areas of the investment. The side walls of the invested flask are hammered first; as an alternative, some companies employ a hydraulic or air ram that pushes the investment from the flask. After that, the investment around the button is hammered until it fractures. Once exposed, the button is gently hammered to send vibrations throughout the casting, which fractures the investment surrounding them.

Combined, these three steps will eliminate about 80 to 90 percent of the investment surrounding the castings. The remaining investment can be removed through water-blasting or soaking in pickle solution, followed by the same ultrasonic and rinsing procedures as in wet breakout. (Bead blasting is sometimes done, but it not only can affect the metal but also pulverizes the investment powder, allowing particles to become airborne. Pickling offers a better, safer alternative.)

Since vibration is an intended result, you should use a metallic hammer for this process. Most casters use a steel hammer, since it transfers the most energy; resinous and non-marring hammers and mallets are not as efficient or effective. One word of caution, though: Be careful that your hammering does not damage the castings. High-volume casters often speed up the

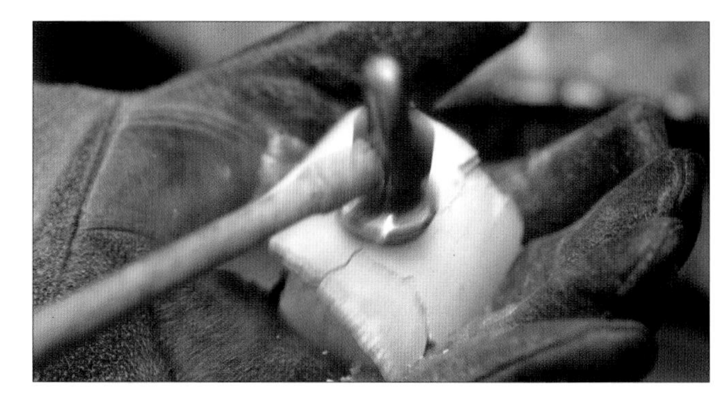

breakout process by using pneumatic hammers, which can often generate sufficient force to distort the castings. These hammers should never be set to their highest impact.

However, even the most careful casters commonly find that a casting has broken off from the tree or base. Consequently, it is a good idea to break out the casting in a contained area. This will allow you to sieve through the investment and recover these separated castings. (You also may find small bits and pieces of metal that surrounded the button during casting.)

## Avoiding Hazards

As I've said, many casters favor wet breakout for the ease and speed it offers. Yet, in addition to offering casters flexibility in deciding when to break out their flasks, dry breakout does offer two important advantages. First, wet breakout poses greater health risks. Steam rising from the quenched flask could contain silica from the investment; if inhaled, silica can lead to silicosis. (It is also a possible carcinogen.) Many casters who are good about wearing respirators when mixing investment think nothing of quenching without proper protective equipment; they feel that, since the investment is now solid, there's little risk involved. This is a mistaken supposition. When quenching, you must always wear a properly rated respirator and have a properly designed ventilation system for hazardous particulate.

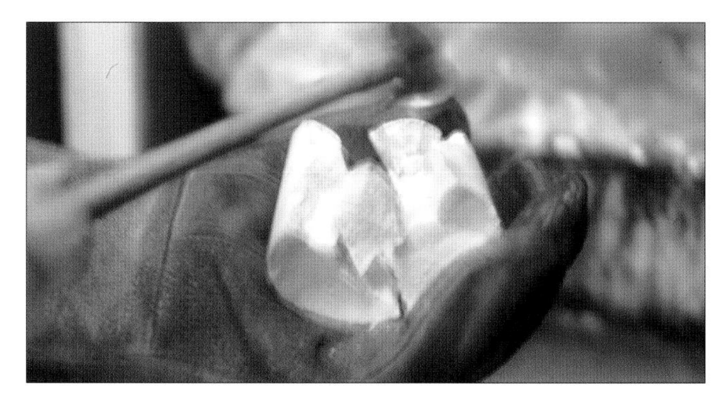

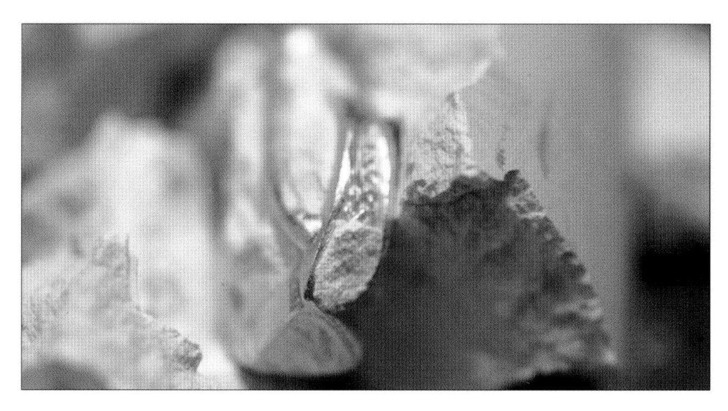

Second, dry investment is considered inert, and thus can be disposed of as a solid waste in most public landfills. (This is fairly common among most states, although it's best to consult your local regulations.) The disposal of wet investment is often a different matter entirely. Wet investment is often viewed as liquid or hazardous waste, and must be disposed of in a specialized waste center. This adds to the caster's costs and must be factored into the equation, although the impact could be different depending on the size of the shop.

For the small shop, the laws still have to be followed, but the wet investment can be allowed to dry through evaporation, so that it becomes dry waste instead of liquid waste. Under no circumstances, however, should the investment be dumped or flushed down a drain, or poured out back behind the shop. Large casting houses and manufacturers generate liquid waste in quantities that do not allow for evaporation to be practical.

Ease, speed, flexibility, safety, disposal—all are important considerations when deciding between wet and dry breakout. But these factors alone will not ensure a successful breakout, nor will either method by itself guarantee success. Instead, casters must turn to one more important consideration: the alloy that's been cast. And for that we need a primer on metallurgy.

## It's All in the Metal

When a metal is melted and cast, it transforms from a liquid to a solid. Throughout this transformation, the metal goes through several changes. First it begins to form crystals and lose its fluidity. Then the crystals begin to solidify into an ordered structure, while the metal continues to give off energy in the form of heat. As the crystals become ordered they take up less space, and the structure shrinks. This entire process happens very quickly.

Yet while we can see the metal change from liquid to solid, there are additional changes that we don't see. As the metal cools, some of its alloys begin to fall out of solution at different rates and segregate along the grain boundaries. The migration of the differing metals happens for a longer period than most people are aware of, at temperatures well below the metal's freezing point. These alloys can begin to form secondary microstructures known as phases, which can cause a dramatic change in the hardness of the metal, depending on the elemental metals from which they were formed.

In yellow gold and palladium-white gold, for instance, copper often forms a secondary phase when it solidifies. But the hardness of the alloyed gold and that of the copper are very similar, so we don't perceive any difference. On the other hand, the difference is quite dramatic in nickel-based white gold. As the metal solidifies and cools, it changes from a single-phase to a double-phase structure. The longer the nickel is allowed to form the second phase, the harder the microstructures become—profoundly changing the hardness of the entire metal. And as the relative hardness increases, so does the brittleness.

These transformations directly impact the breakout process. In yellow gold alloys, the formation of a copper phase has little observable affect on the finished casting, so time is not a critical factor. But the critical temperature considerations of nickel-white gold make it necessary to cool the metal rapidly while it is above 977°F (525°C), in order to prevent the formation of the secondary phase. (This temperature can be higher or lower, depending on the mix and ratio of alloys.)

The metal can also be affected by the investment itself, which provides an insulating thermal blanket around the casting. Most gypsum-based investments transfer heat at a rate of 400°F (204°C) per millimeter per minute. This allows the cavity of the investment mold to remain hot enough so that the metal can fully fill it before solidifying. However, after that happens, the heat then prevents the metal from being able to cool quickly and avoid forming a secondary phase. The recourse here is to break the majority of the investment from the casting and allow rapid cooling to take place.

Two other major concerns among casters are grain size and hot tearing. When it comes to grain growth, they need not worry: Grain growth in the casting process is entirely different than it is for annealing. The amount of grain growth is controlled in large measure by the amount of energy stored in the metal—which, in casting, takes the form of heat. When the metal cools and solidifies, the grain structure forms. However, unlike a fabricated metal, the grain does not grow significantly beyond the initial formation due to the lack of additional stored energy. There is little difference in the relative grain size between a casting that has cooled in the investment and one that has been broken out of the investment while still hot.

Hot tearing is another matter. During breakout, there is a period of time when the grain boundaries of the metal are relatively weak due to the elevated temperatures. That's because when metal cools, its alloys don't solidify uniformly: The higher-melting-point alloys solidify first. If the metal is subjected to vibration stress at this critical time, it can break apart along these weak boundaries—what is known as "hot tearing." To help reduce the chances of this occurring, allow the metal to cool to below 1,100°F (600°C) before beginning the breakout. (This problem is most often seen in the nickel-based white alloys, since they are always broken out hot.)

By understanding your metal and its properties, the last part of the puzzle of breakout will be complete. And with the proper method and procedures in place, you can then look forward to watching the fruits of all your labor—your new castings—emerge from the investment. It almost sounds (dare I say it) easy.

# Casting Resources

# Quality Castings, Superior Service

## ALA Casting Delivers Customer Satisfaction

ALA Casting Co. is a second-generation caster that has successfully served the jewelry industry for almost 60 years. It offers the largest selection of quality castings in the country, most of which are pictured in its six catalogs. Included in the new Volume 6 (mailed in September 2003) are 450 pages of the latest fashions, including the very "hot" right-hand rings and chandelier earrings.

ALA is a very sophisticated trade caster, and also one that delivers individual service. It accepts orders of all sizes: No job is too large or too small. Using personal molds and models from manufacturers, the company produces quality castings with the utmost integrity; this has been proven by the fact that ALA continues to be entrusted by its customers with their exclusive models and molds year after year.

All of ALA's 150,000 styles are available in 10k, 14k, 18k, and platinum, and all are manufactured according to the highest standards. Even the casting of platinum, often considered a difficult metal to work with, has been perfected to an art form by ALA's engineering department.

### State-of-the-Art Facilities

Within the walls of ALA's 50,000-square-foot, three-story headquarters, the company's full-time design staff and model makers are capable of meeting the varied needs of even the most demanding customers. The ALA office and factory are among the largest and most modern in the casting industry. ALA also employs a highly professional, competent, and conscientious staff that is ready and willing to meet the needs of every customer, and ensure complete satisfaction at all times. Our fully computerized system further guarantees speedy and efficient service.

ALA is also dedicated to keeping its customers informed of current trends. To ensure its clients remain at the forefront of jewelry fashion, ALA's monthly flyers offer tremendous opportunities to be first with the best.

When you need styling and quality castings delivered on time, there is only one name: ALA Casting. The company's dedication and drive is perhaps best captured in its mission statement:

"As we look to the future, we pledge to maintain our leadership in the jewelry industry, constantly striving to give you the very best in designs, quality, and service. We wish to express our deepest appreciation to all our loyal customers who are such an important factor to our success. We welcome the opportunity and privilege of serving our old as well as new accounts in the future."

### Ordering Made Easy

Call or fax a request for your customer profile form, and you'll soon become another satisfied customer of the biggest and best, ALA Casting.

ALA's sales team is available to make formal presentations. Please call 1-800-252-5579 and ask for Greg Beuckman. You can also visit the company's Web site at *www.alacasting.com*, where you'll find an easy-to-use search engine that will guide you through all of the many styles available. Find out why the ALA catalogs are called "The Encyclopedia of Jewelry Castings."

**ALA CASTING CO. INC.**

21-21 44th Drive
Long Island City, NY 11101
Phone: 1-800-252-5579
Fax: 718-383-3013
alacasting@aol.com
www.alacasting.com

# Billanti: Leading the Industry in Quality, Service, and Dependability

Forty-eight years ago, three siblings joined forces to build one of the industry's most successful companies. John, Pat, and Joe Billanti started the family institution we all know of today as Billanti Casting Co. Inc. Since then, the next generation of Billantis, Rosemarie and Frank, were born, joined the firm, and are now at the helm of the company today. With the strength of experience behind them, Billanti Casting continues to be strong, eager, fierce, and stable.

The Billantis have always been a sort of enigma to the jewelry industry. Perfectionists, they utilize the utmost of technological advances and the latest materials. Standards are raised for excellence. To a Billanti, casting gold, silver, brass, or bronze is second nature. They need not think or ponder as to how the next challenge needs to be worked out. They have the answer just by looking at the model.

Throughout the years, large jewelry firms, artisans, students, and hobbyists have been bringing their original models to Billanti. The integrity of a client's creation is always held in strict confidence. Models made of wax are the substance of choice, but they do not limit themselves and, far be it, limit their clients to this. They have accepted pieces made of wood, flower petals, sea shells, leaves, fired clay, stones, plastics, and even metals for either one-of-a-kind pieces or multiple reproductions involving the manufacture of molds: vulcanized rubber, non-shrink rubber, and even zero-shrink rubber. "We do have size limitations of 4.5 inches by 6.75 inches, though we have been able to accommodate larger items on occasion," they add. No task is too large or too small.

The possibilities are apparently endless.

At their factory in New Hyde Park, New York, the brother-sister team behind the scenes works diligently with a highly qualified staff to make every dream a reality. The pride they take in their work can be seen in the client's final product.

A production house of small intricate sub-factories, with fire blazing and steam pumping. The glory of man's ability to control the elements. The glory of man's ability to soil his bare hands. The Billantis live it every day. They love it. It's bigger than they are. It's in their genes.

Billanti Casting, now almost half a century old, is without a doubt providing the quality and service on which the industry has come to depend.

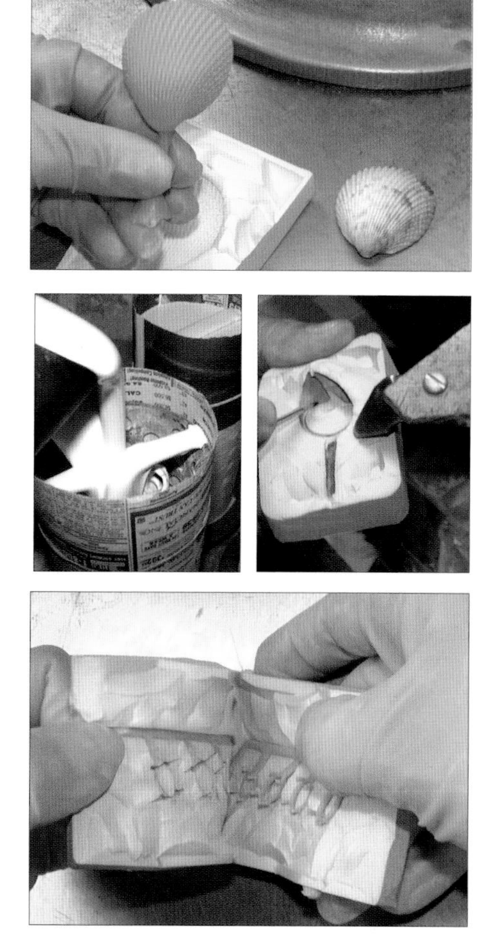

299 South 11th St.
P.O. Box 1117 - Dept. AJM
New Hyde Park, NY 11040
Phone: 1-516-775-4800
www.billanticasting.com

# Lost-Wax Casting with Overpressure; Centrifugal Casting Is Obsolete

## Casting Platinum

Using Eisinger Enterprises' Unique Spruing Method and overpressure casting with the use of high frequency induction melting technology will result in higher detailed, smoother castings. It is the preferred process by leading platinum casting houses around the world.

First and foremost you must start off with good, clean wax or plastic models along with advanced spruing techniques. Technology cannot correct poor quality models or improper spruing.

Eisinger Enterprises' Unique Spruing Method requires less metal for the button than conventional spruing techniques (30-40%) and, combined with the low rotation of the flask, diverts the turbulence of the molten metal to allow the gases to escape from the flask with less resistance than conventional methods. This reduces the chance of trapped gases creating porosity.

Centrifugal casting creates a strong turbulence, which of course can result in porosity and other casting problems.

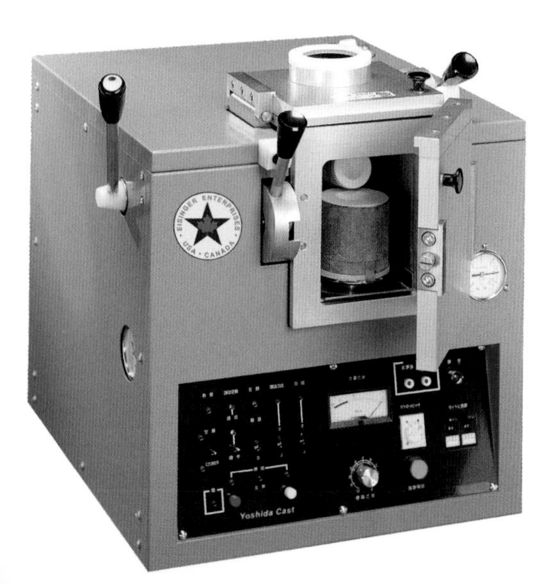

High frequency induction melting provides a medium stirring action ideal for casting platinum. This, combined with a protective gas blanket (argon, nitrogen, or forming gas) on top of the molten metal, will reduce the oxygen in and around the metal, therefore reducing the discoloration (oxidation) of the metal when it is cast.

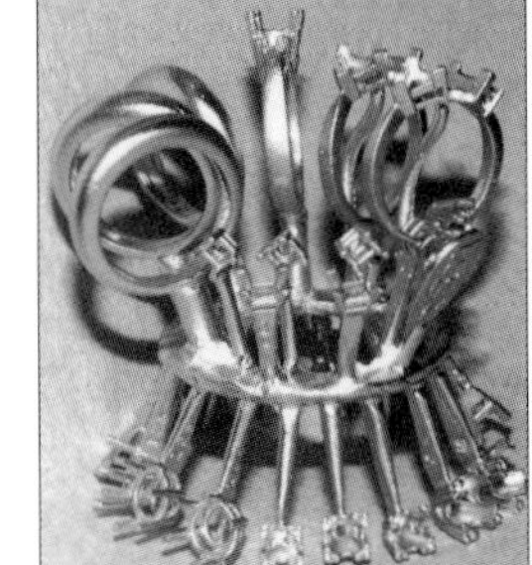

With the combined active pressure of 4.8 bars (70.7 psi) and reduced turbulence, you achieve denser castings. With this combination, Eisinger Enterprises and Yoshida make platinum casting more affordable by reducing your rejection rate, increasing productivity and increasing the density of your casting vs. conventional casting techniques.

## Casting Gold and Silver

Eisinger Enterprises recommends the use of medium frequency induction melting technology. This provides a light stirring action ideal for casting gold and silver. This, combined with a protective gas blanket (argon, nitrogen, or forming gas) on top of the molten metal, will reduce the oxygen in and around the metal, therefore reducing the discoloration (oxidation) of the metal when it is cast.

When using perforated flasks in a sealed vacuum assist casting chamber, you are able to distribute the negative pressure (1 bar) evenly throughout the chamber, in essence pulling the metal down into the cavity. The size of the casting machine and the flask, as well as pattern characteristics, will determine the amount of overpressure that is needed; an average 4" x 8.5" flask for gold requires a minimum of 1 bar. As you can see, this combination of vacuum and overpressure doubles the active pressure forcing the metal into the flask cavity. With all of these features combined, you can lower your flask and metal casting temperature for gem-quality stone-in-place casting and cast finer detailed pieces, produce denser casting, reduce the chances of porosity, and decrease your cleanup time. Eisinger Enterprises and Vetter Technik offer three systems with these features... Which one is suited for you?

Quick references when we refer to:

Vacuum (negative pressure): 1 bar or 29.99" of mercury or 14.7 psi at sea level pulling the metal into the flask cavity.

Overpressure (positive pressure): 1 bar or 14.7 psi, 2 bars or 29.4 psi, 3.8 bars or 55.9 psi pushing the metal into the flask cavity.

Active pressure is the combined force of vacuum and overpressure.

242 Astor St.
Newark, NJ 07114
Phone: 1-800-282-1980
Fax: 973-596-1329
eisinger@earthlink.net
www.eisinger.com

# Consistent Casting Alloys

## David H. Fell & Co. Helps You Ensure Quality

Since 1973, David H. Fell & Co. Inc. has provided the highest quality precious metal products and refining services to the jewelry industry. It is a family-run business created to reflect the ideals of their founder, David Fell: high quality, personal attention, and honesty. With these ideals, DHF Co. has continually developed casting alloys that add variety to their metal selection, and inspire new ideas for manufacturers and designers.

### Karat Gold Casting Alloys

DHF Co.'s karat gold casting alloys are offered in a wide spectrum of colors in 10k to 22k. There are several whites and shades of yellow, as well as red and green. The vast selection includes DHF Plus alloys, which brighten castings and minimize cleanup, and also the standard alloys that customers have relied on since the company's inception. Many casting alloys have a grain refiner that produces a finer grain structure and increases reusability.

"David H. Fell & Co. Inc. continually develops new products that add variety to our product line," says Ruth Fell Failer (below), VP of Sales and Marketing. Recently, two new white golds have been added to DHF Co.'s extensive list of casting alloys: DHF Satin White and DHF Polar White. DHF Satin is very malleable, making it an excellent choice for pavé work. It has a good reusability ratio. The DHF Polar White also has great workability and is whiter than the Satin White alloy. It may not need to be rhodium plated.

Custom colors are available and DHF Co. can manufacture to most customers' specifications.

### Platinum Casting Alloys

Platinum customers have a diverse selection of casting grains available from David H. Fell & Co. The platinum alloy selection consists of pure platinum, iridium platinum, ruthenium platinum, and cobalt platinum. As with their gold alloys, DHF Co. will manufacture many custom platinum alloys to meet the buyer's needs.

### Silver Casting Alloys

Fine silver and sterling silver grains complete the variety of casting alloys from DHF Co. As it does with the gold alloys, DHF Co. manufactures a sterling silver alloy in a "Plus" formula. Again, the "Plus" is a grain refiner that reduces the size of the grains to help the metal fill in fine details.

Whether a customer is casting gold, platinum, or silver, David H. Fell & Co. has an alloy to suit their every need. A free DHF Products and Services Catalog is available by calling 1-800-822-1996.

The catalog contains several pages of useful casting hints and tips. Customers may also visit our Web site at *www.dhfco.com*. Here there are many useful casting hints, as well as some interactive conversion charts to assist in converting wax to metal weight or reducing and raising the karats of a particular alloy.

David H. Fell & Company, Inc.
Precious Metals Refiner and Manufacturer
of Quality Mill Products

P.O. Box 910952
Los Angeles, CA 90091-0952
Phone: 1-800-822-1996
Fax: 323-722-6567
info@dhfco.com
www.dhfco.com

# For Wax Carving and More

## Otto Frei Lets You Do Your Best Work

Where does quality jewelry start? Some would say it's in the selection of the best metals. Others might suggest that it's all a function of a well-thought-out design. Still others might offer that quality stems from the work process itself. Ask Steven Frei, president of Otto Frei in Oakland, California, and he'll tell you that quality starts with the bench you work on—and the best benches in the jewelry business come from Otto Frei.

"The John Frei Custom Work Bench is the best quality bench jewelers can buy," Frei says. "When we first came out with this bench 20 years ago, we weren't sure we could sell it because it was, and still is, more expensive than other benches. But we found that jewelers were thrilled to have a bench of this quality.

"Some of the largest jewelry manufacturing shops in the United States have outfitted their jewelers with our benches," Frei adds. "They want to show their jewelers that they're appreciated, so they give them the best workstations."

The John Frei Custom Work Bench is built in cabinet-quality solid maple or oak with a 2-inch-thick butcher-block top, and it comes fully assembled. The standard bench is 49 inches wide—almost a foot wider than most other commercially available benches. Choose from six different stock models or have the company create a custom bench to suit your needs. Whatever size, shape, wood, color, or options you want, Otto Frei can make it for you. When a customer doing remount work recently needed a bench on wheels, he got it from Otto Frei.

Once you've treated yourself to the best bench, you'll want to be sure it's equipped with the best tools—and you'll find them all with Otto Frei. "We're known for carrying the highest-quality tools and equipment from top manufacturers around the world," Frei says. "And we're always on the lookout for more." Recent additions include Otto Frei's Setter's Microscope, as well as new disc mass-finishing systems.

"Although we offer all the classic hand-finishing tools and the best polishing buffs and compounds, these disc finishers are saving jewelers a lot of labor," Frei says. "You can get hand-polishing-quality results in under an hour."

Otto Frei is also the place to go for your platinum jewelry needs, offering the best in platinumsmithing tools, including the widest selection of platinum polishing products on the market. "We have products from Germany, Japan, Italy, and the United States," Frei says. "People have their preferences, and we let them decide what they want."

If there's one thing everyone wants, it's quality, and that's why so many jewelers turn to Otto Frei. From the products on the bench right down to the bench itself, if you want to do your best work, work with Otto Frei.

126 Second St.
Oakland, CA 94607
Phone: 1-800-772-3456
Fax: 800-900-3734
sales@ottofrei.com
www.ottofrei.com

# Let Your Creativity Soar

## In GIA's New Applied Jewelry Arts Program

Whether your goal is to become a jewelry designer, jewelry production manager, manufacturing executive, quality control professional, or jewelry business owner, the skills and training you'll acquire at GIA will give you solid footing on your career path.

GIA's new six-month Applied Jewelry Arts diploma program teaches you the fundamental skills you need to begin establishing yourself as a true jewelry design and manufacturing professional. GIA's world-renowned instructors will show you how to tap into your creativity and transform your ideas into beautiful pieces of jewelry. You'll learn skills in design, computer modeling and rendering, wax carving, mold making, and casting—the crucial first steps in the creation of your jewelry.

### Program Highlights

**Jewelry Design**—Understand jewelry design theory and learn how to communicate design ideas to customers and manufacturers with professional illustrations of jewelry and gemstone shapes, forms, and textures.

**Introduction to CAD/CAM, Intermediate CAD/CAM, and Advanced CAD/CAM**—Look at your designs from every angle in 3-D and learn to create accurate and precise models for virtually all types of jewelry, including pendants, earrings, and rings. Learn to use 3-D software to design rings, heads, bezels, gemstones, diamonds, and channels. And learn to take a design from creation to rendering and finally to wax.

**Comprehensive Wax Techniques**—Create a wide variety of designs in wax, including additive and subtractive methods; how to hollow models in preparation for casting; and create mirror-image pieces.

**Casting**—Learn horizontal centrifugal casting, as well as tabletop vacuum and chamber vacuum methods. Cast various types of jewelry, from basic free-form shapes to complex traditional and contemporary jewelry. Become proficient in casting techniques, including spruing, treeing, and casting stones in place. Spend time troubleshooting common casting problems and learn the unique considerations involved in platinum casting.

**Mold Making**—Explore the art of vulcanized rubber and room temperature vulcanizing (RTV), and learn how to avoid common and costly pitfalls in mold making and cutting techniques. Learn the entire mold making process from start to finish.

When you complete GIA's new Applied Jewelry Arts (A.J.A.) Program, you have more than a mere understanding of the jewelry pre-production process. What you learn will prove to be both unforgettable and invaluable. It will

stimulate your creativity, test your ability, and ultimately seal your position as a true professional in the jewelry industry.

### About GIA

Founded in 1931, the nonprofit Gemological Institute of America is internationally recognized as the world's foremost authority in gemology. The Institute translates its knowledge of diamonds and colored stones into the most respected gem and jewelry education available anywhere.

To learn more about GIA's jewelry manufacturing programs and courses and all the gemology, sales, and jewelry business courses GIA has to offer, please call today for a free course catalog: 1-800-421-7250, ext. 4001, or visit *www.gia.edu*.

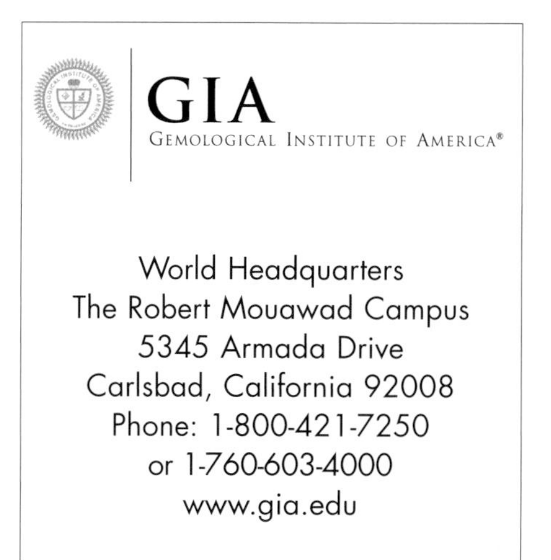

# Bread & Butter

## Wax and Investments for Every Need

There have been many advancements in products and technology in the field of lost-wax casting, but the one thing that has not changed is the need for a consistent high-quality injection wax and investment. Lucky for Kerr, we produce the best of both.

KerrLab's Accu® Flakes™ offer the lowest ash content available (0.003%), guaranteeing the cleanest burnout possible. Available in seven application-specific colors in quick-melting flake form: Aqua Green, Ruby Red, Turquoise Blue for high detail, NYC Pink for filigree, Flex Plast Blue for flexibility, Tuffy Green for large, flat patterns, and Super Pink for bezels.

Our Satin Cast Line features only USA "Zero Mineralized" cristobalite. Satin Cast Original is the original Satin Cast Formula with over 50 years of service to the jewelry industry. Satin Cast 20 is KerrLab's premier casting investment—often called "the most forgiving" investment. Satin Cast Xtreme offers the highest cristobalite content, giving it a high heat tolerance for many 18 karat white metals. Satin Cast Diamante is the newest additon to the line, offering all of the benefits of Satin Cast 20, yet formulated specifically for stone-in-place casting.

For more information, contact your authorized KerrLab distributor, or visit us online at *www.kerrlab.com*.

1717 W. Collins Ave.
Orange, CA 92867
Phone: 1-714-516-7650
Fax: 714-516-7649
kerrteam@kerrlab.com
www.kerrlab.com

# The Changing Role of the Contract Caster

More than ever before, the contract caster is being called upon to be a one-stop source for all types of jewelry-making operations. It is no longer enough just to provide quality castings at a reasonable price. Today's casting house must be a virtual supermarket of services.

From model making to finishing, savvy customers are outsourcing their production needs to the caster. Why go to a caster? Because he is right in the middle of the production cycle, and so is in a unique position to observe and influence every step of the process and apply his expertise to creating the best piece of jewelry with the least number of headaches for the client. The caster is not then merely a caster; he is a solution provider.

So what distinguishes one caster from another? How should a customer choose between the growing legions of contract casters offering their services? There are several qualities that one should look for in a casting house: experience and reliability, a range of services offered, flexibility, and innovation.

Quality Casting Inc. has been furnishing the jewelry industry with fine quality castings for 20 years. From the largest manufacturer to the designer to the one-person repair shop, they treat each customer with individual care and attention. Each job, large or small, is treated with equal respect and importance. And Quality is well known for their superior ability to handle special orders. Their information management, organization, and computerized order tracking systems allow them to give this specialized attention, which is unparalleled in the casting business.

"I was ready to have to fight for my work. I thought that I would be too small a user to get the kind of attention that I wanted," says Cynthia Nockold, a New York-based designer and model maker. "I was honestly surprised at the support and the services that Quality had to offer."

Quality never stops striving to refine their level of consistency, value, and service, day in and day out. They continually upgrade their equipment and, even more important, their procedures and process controls, to remain at the forefront of the casting industry.

Owner and company president Carl Morfino reminds us: "Good customer service and sound technical advice are every bit as important as good quality product. I know other companies can cast platinum, gold, and silver. That's no secret. We feel that's not enough anymore. Our customers are demanding more. What sets us apart is that we can receive, track, and present information to our customers in a useful fashion that most other companies can't. We are constantly revising our systems to accommodate our customers' needs."

General Manager Frank Reisser adds: "Customers today want their work quickly, of course, but they also need instant information. They need to know when and how they will be getting their castings so they can schedule their time appropriately. And many of them require the expanded range of services we offer. We are providing model making, stone-in-place casting, and technical advice beyond casting questions; we're even providing finishing options for our customers' convenience."

The Quality Casting formula. Providing solutions, information, and services in a single source for the needs of the jewelry industry.

# Successful Stone-in-Place Casting Is Available to All

Manufactured of .925 silver, ClickSet™ master model-making components are easily shaped and manipulated by the model-maker, yet contain the internal mechanics for successful, repeatable stone-in-wax casting results. All the necessary seat grooves have been precisely calibrated and are pre-located in ClickSet components. Engineered to high tolerances, ClickSet components account for all manufacturing shrinkages and each has been fully tested on a wide variety of alloys to ensure that stones will be secure after casting.

With ClickSet, you bypass the calculations, labor, and time necessary to create and test the stone-in-wax settings you want. Instead, you can concentrate on designing your master models with complete confidence that the stones you cast in place will fit perfectly the first time, every time!

First, choose the ClickSet master(s) you need to create your design. Then alter, modify, or shape the external surfaces of the metal master to suit your design vision. Changes to the exterior of the components will not affect the stone-in-wax setting properties. Be extremely cautious when working near the inside surfaces, however; do not change anything inside the setting where the stone will be seated.

**To Ensure the Best Results:**

- Use high-quality, calibrated stones with consistent size and cut. ClickSet settings are engineered to close tolerances and will not accept large variations in stone size.
- Use a high-quality wax with good flexibility and memory. After you click stones in place, you want the wax to snap back into its original shape.
- Account for the depth of your stone—the culet of the stone should seat within the model's depth.
- Master models must have an open back. After burnout, enough investment must be left behind to support the stones from at least two directions.
- Allow for the standard 4–6 percent shrinkage of overall thickness during molding, waxing, and casting.
- Make ring masters 1–1.5 sizes larger than the finished size you want. *Important:* Never hammer ClickSet models; this can severely distort their precision internal mechanics.
- Pre-polish all recessed areas of your master.
- Shape the feed sprue for your master to ensure that both wax injection and metal flow during casting will be as smooth as possible.
- After casting, flasks must cool enough to be comfortably held in your bare hand before breaking out to avoid cracking your stones.

**Benefit From:**

Quick Start

Start stone-in-wax setting very quickly with a minimum investment in time or personnel.

Complete Compatibility

Use ClickSet settings as side or accent stones with regular settings.

Better Settings

High quality and precision engineering ensure a perfect stone-setting and minimal finishing time.

Perfect Fit

With molding and metal casting shrinkages pre-calculated, calibrated stones fit perfectly.

Fast, High-Quality Results

Simply "click" your stones into place in the injected waxes for fast, high-quality results.

Extended Markets

Offer customers new designs at affordable prices, giving you new markets and higher profits.

7500 Bluewater Road NW
Albuquerque, NM 87121
Phone: 1-800-545-6566
Fax: 800-965-23291
info@riogrande.com
www.riogrande.com

# Sierra Pacific Casting Inc.

## Quality, Confidentiality...From Start to Finish

High quality casting and finishing in gold, silver, fine silver, and bronze has been the hallmark of Sierra Pacific Casting since 1974.

Founded in San Francisco, we moved to our new, larger facilities across San Francisco Bay in 1999, where we have continued to serve an ever-growing client base of designers, individual craftspeople, distributors, and other manufacturers. Goods produced in our facility are sold in a variety of shops, from retail chains and small boutiques, as well as through catalogue and online sales.

Our new production facility is equipped with the latest technology developed for our industry. With induction atmosphere- and temperature-controlled casting systems and automatic wax injection machines, Sierra Pacific Casting's goal is to produce the highest quality product at the most competitive price and to provide unsurpassed service to all our clients, regardless of whether they are ordering one piece or thousands.

We produce a variety of molds for production purposes, including vulcanized rubber molds, RTV (room temperature vulcanization or cold molds), and larger, circular wax spinning molds. Each type of mold provides a particular benefit, and its use and application is discussed with each client. Molds can be stored in our facility or can be returned to the client for storage in his/her facility. All our work is strictly confidential. Sierra Pacific Casting does not produce its own product lines; we are strictly a contract manufacturer.

Our finishing department creates a range of finishes for items ordered in production quantities. Small silver and gold goods may be antiqued and tumbled in a centrifugal barrel finisher and given a final finish with porcelain media. This produces a bright, polished product with a darkened background. We often use a combination of machine and hand finishing to finish other goods which may need a more hands-on approach. We also can fuse or solder findings such as posts, tacks, joints, and catches.

Ordered items are tracked in our computer system from initial order entry to final production and shipping. Each production item is assigned a casting number and entered with pricing and, if

appropriate, quantity discounts. After casting and/or finishing, the castings are sorted by item number and then inspected. Once they have successfully passed our scrutiny, they are counted and invoiced. Our invoices are easy to read, and one of our office staff is always available to answer any questions.

Sierra Pacific Casting takes great pride in our skilled staff of wax injectors, mold makers, casters, finishers, and other personnel. We provide all our staff with competitive salaries and benefits and comply with all labor, environmental, and other regulations affecting manufacturing and, in particular, manufacturing for the jewelry industry.

292 Fourth St.
Oakland, CA 94607-4332
Phone: 1-510-444-0550
Fax: 510-444-0208
info@sierrapacificcasting.com
www.sierrapacificcasting.com

# Solidscape Introduces New Models for Easier Casting

Solidscape Inc. designs, develops, manufactures, and sells model making machines and software that are used in the production of fine jewelry. Manufacturing is performed at Solidscape's headquarters in Merrimack, New Hampshire (USA). Progressive jewelers throughout the world have purchased over 700 Solidscape systems starting in 1995.

The Solidscape T66 and the new T612 are used in conjunction with 3-dimensional jewelry design software to automatically grow physical models of jewelry designs that can be directly investment cast. Solidscape's systems are compatible with all of the popular 3-D jewelry design packages currently on the market.

The T66 and T612 systems jet a proprietary thermoplastic (ProtoBuild™) for growing models and a dissolvable wax (ProtoSupport™) support structure to support cavities, overhangs, and undercuts. By using the dissolvable support structure, fine feature detail can be achieved without fear of breakage that can occur with other technologies. In addition, a dissolvable support structure provides a high-quality surface finish because nothing has to be broken, cut, or scraped away. The ProtoBuild™ material has been formulated specifically for investment casting applications and exhibits no shrinkage and a clean, low temperature burnout. Both the model material (ProtoBuild™) and the dissolvable support material (ProtoSupport™) are non-toxic and inexpensive to use.

The T66 and T612 operate unattended so that models can be grown overnight or over the weekend, thus maximizing throughput. During the day personnel can concentrate on establishing new business and preparing designs to be grown. In the morning, designs grown overnight can have the support structure dissolved in approximately 15 to 30 minutes and the models are ready to be cast. Typically, three to four designs can be produced overnight.

Models are grown in layers. Each layer is built on top of the previous layer until the design is completely grown. After each layer is grown, the surface of that layer is "shaved" flat by a rotary blade device, and shaved material is vacuumed away. This process ensures high accuracy and surface finish. Design characteristics determine the optimum user-selectable layer thickness with which to grow. Because the material used to grow models exhibits no shrinkage in investment casting, the shrinkage factor of the metal-of-choice is specified during job setup, and models are automatically sized to accommodate casting. This is very important when considering designs such as invisible and pavé settings, where sizing of the cast design can be a challenge.

Because production demands vary between establishments such as a custom jeweler and a manufacturer, Solidscape markets two systems that vary in build capacity. The T66 has a build area of 36 inches (91.4 cm). This means that a single job might include several rings, a pendant, a bracelet, and earrings. The new T612 model offers all of the features of the T66 but has a build platform of 72 inches (182.8 cm). In addition, the T612 operates at a speed 25% greater than that of the T66.

Solidscape recently opened a sales and service office in Los Angeles, California (USA), and established its European headquarters in Utrecht, Netherlands, to better serve these markets.

316 Daniel Webster Hwy.
Merrimack, NH 03054
Phone: 1-603-429-9700
Fax: 603-424-1850
precision@solid-scape.com
www.solid-scape.com

# Stebgo Metals Inc.

## Specialists in Quality Casting Grain

Stebgo Metals Inc. is a major supplier of casting grain in the jewelry industry. Stebgo manufactures a variety of gold casting grain and supplies silver, platinum, and other alloys for your casting needs. At Stebgo, our casting grain has specialized properties that improve castability, reduce rejection rates, and produce a superior quality finish.

Stebgo Metals provides casting grain for all of your casting needs. Our recommended grain for most gold casting is made with special de-oxidizers. These elements of boron and silica give this grain excellent reusability, high tarnish resistance, and the elimination of fire scale. We also provide nickel, nickel-free, and palladium white gold casting grain. Stebgo manufactures a variety of colors, including peach, pink, and green gold, for all of your specialty projects.

Stebgo specializes in rolling and fabricating casting grain. These casting grains have little or no de-oxidizers and are formulated with grain refiners for optimum grain size, giving maximum efficiency in rolling and wire drawing application.

**Casting Tips & Issues**

The cause of most casting rejection is porosity. Porosity can be greatly minimized when all stages of the casting process are done with great care and attention to detail. The major causes of porosity are burnout problems, improper spruing, improper flask temperature, the style of the piece, and problems related to non-deox alloys.

Proper burnout cycles are paramount to having an excellent cast. Be sure to have smooth temperature transitions during the program cycle, and check to make sure all carbon residues in the flask are eliminated. This may involve a longer burnout cycle and leaving it at the highest setting (generally 1,350°F) for a longer period of time.

Proper mixing of the investment powder is critical. Follow your investment manufacturer's guidelines. Make sure your investment has not separated or gone past its shelf life, and keep your investment free from moisture. Make sure to wear a particle mask when investing.

When spruing your pieces, always try to visualize the flow of the gold into the cavity of your flask. Eliminate, if possible, any sharp angles and areas that pinch off the flow of metal. When the sprue is too constricted you may get the gold spraying into the cavity, causing fine porosity and "hot spot porosity" where the sprue connects to the piece.

At Stebgo, we have flask temperature recommendations for each metal alloy. Improper flask temperatures may cause porosity or an incomplete cast. The most common flask temperature for gold is around 1,000°F; depending on the type of alloy used and size of the piece, this temperature may need to be raised or lowered.

Stebgo offers free technical service to help you with troubleshooting problems encountered in regular jewelry manufacturing operations. We are committed to making the casting process a success for you and your operation. Please ask for our free catalogs that show the rest of our product line and services, including refining, tools, findings, investment, gold, silver, and platinum products.

*Stebgo Metals, Inc.*

P.O. Box 7
South St. Paul, MN 55075
Phone: 1-800-289-0138
or 1-651-451-8888
Fax: 651-451-1397
info@stebgo.com
www.stebgo.com

# Gold Jewelry Casting United Precious Metal Refining

United Precious Metal Refining Inc. is a major supplier of master alloys for the precious metals industry, with sales offices throughout the world. United manufactures a variety of master alloys for gold, silver, and platinum. With ongoing research, United is in the forefront of developing new alloys with specialized properties that improve castability, reduce rejection rates, and produce a superior quality finish.

### De-Oxidized Alloys

Alloys are formulated with or without special de-oxidizers, which are carefully controlled to give optimum results. Some of the advantages of using alloys with de-oxidizers are excellent metal reusability, high tarnish resistance, and the elimination of fire scale. Raw materials in United alloys are oxygen free and must pass stringent quality controls. Nickel, nickel-free, and palladium white gold alloys are available. All white gold alloys containing nickel are tested to ensure proper compliance with the European Regulation EN 1981 for nickel release.

### Non De-oxidized Alloys

United also specializes in rolling and fabricating alloys for use in wire drawing, chain making, and rolling sheet. These alloys often have little or no de-oxidizers and are formulated with grain refiners for optimum grain size to give maximum efficiency in rolling and wire drawing applications. United also manufactures custom specialized alloys at the customer's request.

*United's vast assortment of yellow, white, pink, and green alloys can accommodate all jewelry manufacturers' needs.*

### Porosity

The subject of porosity in jewelry investment casting is a complex issue. We must deal with many interrelated factors to minimize porosity rejection rates and increase quality. The major reasons for porosity are burnout problems, improper spruing, flask temperature, style of piece, and the type of alloy used, de-oxidized vs. non de-oxidized.

### Investments

Proper mixing of the investment powder is critical in maintaining a quality cast. The only procedures to follow are the manufacturer's.

### Burnout

Proper burnout of wax patterns is important to a quality cast. Many casters use a programmed burnout cycle which allows a smooth transition of wax evacuation without leaving carbon residue in the flask.

### Technical Service

United offers free technical service to help you with troubleshooting problems encountered in regular jewelry manufacturing operations. United's expertise in production casting is unparalleled in the jewelry industry. To contact our technical hotline, please call 1-800-999-FINE or "Ask Doc for technical help" at our Web site, *www.unitedpmr.com*. E-mail *doc@unitedpmr.com*.

2781 Townline Road
Alden, NY 14004
Phone: 1-800-999-FINE
or 1-716-683-8334
Fax: 716-683-5433
sales@unitedpmr.com
www.unitedpmr.com

# Ambition Fuels World Recognition of Zero-D Products

Zero-D Products Inc. was established by Bill Mull in 1984 for the purpose of brokering, as well as manufacturing, machined plastics and molded rubber parts for industry. During the early years, exposure to and contacts within the precious metal jewelry industry identified a need for higher quality mold rubber products used in the casting process. This new focus led us to develop first our natural rubber product—Akron Jewelry Rubber—which was introduced in 1991.

Heartened by the excellent reception Akron Jewelry Rubber received from the industry, we began development of a series of silicone mold rubbers, led by our 4X Gold, which had far higher tear resistance than any other product on the market at that time. Several other attributes, such as extreme high wax finish and easy release properties, made 4X Gold a popular alternative to the less sophisticated formulas available at the time.

In the first few years of mold rubber production, the milling and calendering of our rubber products were subcontracted to some of the fine rubber manufacturing houses that populate the greater Akron area. As order volumes expanded, we purchased the necessary rubber manufacturing equipment to allow us to handle all aspects of production in-house. This provided us with the ability to monitor the quality of our products even more closely, and to respond to customer requests in a much more timely manner.

The next logical product line extension came in the development of a clear liquid silicone RTV. Our objective again was to provide a product that offered the highest tear strength available in order to better serve the needs of the manufacturing jeweler. After

nearly a year of rigorous developmental lab work, Akron RTV Silicone was ready for the market. Today, it is recognized around the world as the most dependable cold-cure system for a variety of applications. Unlike the less costly urethane mold rubbers, Akron RTV Silicone molds are permanent. They do not degrade over time due to contact with moist conditions and are not subject to urethane's natural tendency to revert to a sticky state.

Heightened awareness of our "Zero Defects" philosophy led to more and more requests for new products from Zero-D. With the same careful consideration to quality, we added silicone casting gaskets, Slick Silicone Mold Release Spray, and Poly Set two-part mold putty.

Our most ambitious undertaking in the last four years was the development of our Akrovest Premium investment and Polyhedral injection waxes, which are winning wide acceptance for quality and value. These superior grade products fully complement our mold rubber offerings, and collectively provide the casting jeweler with a one-stop source for the high-quality materials required to produce high-quality precious jewelry.

Recently we introduced Akrovest Platinum II investment. This is a stunning contribution to improvements in efficiency and safety for the industry. Platinum II is the very first safe and dependable, water-only platinum investment to be available. The elimination of acid from the mix satisfies old safety issues, and averts the very high cost of shipping acid. With acid and phosphate investments, breakout is very difficult. With Platinum II, a simple light tap will powder the investment. With care, water quench also works well.

As we move forward, our pledge of Zero-D—that is, zero defects—is foremost in our thinking, planning, research, and development. Over the last 11 years of production, as hundreds of thousands of pounds of rubber have passed through our mills, Akron Jewelry Rubber has maintained a flawless record of quality and consistency. We are very proud that our products have been accepted as the quality benchmark against which others are judged.

Today, Zero-D Products Inc. markets its mold rubbers, investment, and casting waxes throughout the United States and in more than 50 countries around the world. A full list of the products available can be viewed on our Web site: *www.zerodproducts.com.*

38285 North Lane, #103
Willoughby, OH 44094
Phone: 1-800-382-3229
or 1-440-942-1150
Fax: 440-942-2130
jwryrubber@aol.com
www.zerodproducts.com

# About the Contributors

**Daniel Ballard** is the national sales manager for Los Angeles-based Precious Metals West/Fine Gold.

**Eddie Bell** is president of Neutec in Albuquerque, New Mexico, and one of the founders of the annual Santa Fe Symposium on Jewelry Manufacturing Technology.

**Michael Bondanza**, president of Michael Bondanza Inc. in New York City, has received numerous design awards for his creative use of platinum.

**Christopher J. Cart** is vice president of manufacturing and marketing at Au Enterprises in Berkley, Michigan.

**Paul Coen** has served the casting industry for more than 30 years. He has held prominent positions with both Kerr and Hoben David Ltd.

**Elaine Corwin** is vice president of technical services for Gesswein in Brockport, Connecticut. She can be reached at 1-800-544-2043, *ecorwin@gesswein.com.*

**Gary Dawson** is the owner of Goldworks in Eugene, Oregon.

**Stewart Grice** (BSc CEng MIM) is mill products director for Hoover & Strong Inc. in Richmond, Virginia.

**Greg Gilman** is tools and supplies specialist with Stuller Inc. in Lafayette, Louisiana. He is also, with Gregg Todd, the co-author of *At the Bench: An Illustrated Guide to Working with Gold and Silver* (MJSA/AJM Press, 2002).

**Chuck Hunner** operates a studio in the Blue Ridge Mountains of North Carolina, making "jewelry that touches the heart." He can be reached at *chuck@golden spirit.com.*

**Charles Lewton-Brain** is an educator, goldsmith, author, and proprietor of Brain Press in Calgary, Alberta, Canada.

**Jurgen J. Maerz** is director of technical education for Platinum Guild International USA and the author of *The Platinum Bench: An Illustrated Guide to Easy Platinum Repairs and Fabrication* (MJSA/AJM Press, 2001).

**John McCloskey** is executive director of metals for Stuller Inc. in Lafayette, Louisiana.

**Marc "Doc" Robinson** is technical director with ABI Precious Metals in Carson, California. He has written many articles for *AJM*, as well as the In The Shop column. His e-mail address is *goldmanic@tazland.net.*

**Bob Romanoff** is the president of Romanoff International Supply Corp. in Amityville, New York. He can be reached at 1-631-842-2400, *romanoff8@aol.com.*

**Timo J. Santala** (1961-2001) was a manufacturing expert, writer, and lecturer who, at the time of his death, was president of the Jacmel-owned Soloro Manufacturing Corp. in the Dominican Republic.

**John Sciarrino** is the owner of Giovanni Photography in Naples, Florida. He frequently photographs the work of area jewelers.

**J. Tyler Teague** of JETT Research in Albuquerque, New Mexico, is a global jewelry industry consultant who specializes in factory design and process engineering. He can be contacted at *tyler@jettresearch.com.*

**Gregg Todd** is industrial training and project administrator for Stuller Inc. in Lafayette, Louisiana. He is also, with Greg Gilman, the co-author of *At the Bench: An Illustrated Guide to Working with Gold and Silver* (MJSA/AJM Press, 2002).

**Suzanne Wade** is a contributing editor to *AJM* and the magazine's former associate publisher/editor.